OPTICAL TRANSDUCERS AND TECHNIQUES IN ENGINEERING MEASUREMENT

OPTICAL TRANSDUCERS AND TECHNIQUES IN ENGINEERING MEASUREMENT

Edited by

A. R. LUXMOORE

*Department of Civil Engineering,
University College, Swansea, UK*

APPLIED SCIENCE PUBLISHERS
LONDON and NEW YORK

APPLIED SCIENCE PUBLISHERS LTD
Ripple Road, Barking, Essex, England

Sole Distributor in the USA and Canada
ELSEVIER SCIENCE PUBLISHING CO., INC.
52 Vanderbilt Avenue, New York, NY 10017, USA

British Library Cataloguing in Publication Data

Optical transducers and techniques in engineering
measurement.
1. Measuring instruments   2. Transducers, Optical
I. Luxmoore, A. R.
621.37′9   QC100.5

ISBN 0-85334-203-2

WITH 162 ILLUSTRATIONS

© APPLIED SCIENCE PUBLISHERS LTD 1983

© CROWN COPYRIGHT 1983—Chapter 5

D

621. 379

OPT

Printed in Great Britain by Galliard (Printers) Ltd, Great Yarmouth

PREFACE

Optical measurement techniques have been stimulated in recent years by the advent of lasers and also by modern electro-optical devices. Despite the considerable research and developments in this field, these techniques are not widely appreciated by engineers, who are often unaware of their versatility. This book provides a single comprehensive source giving the basic science and technology involved in the implementation of these latest methods, for use by industrial and research engineers, in the solution of measurement problems and the design of measurement systems. The book covers the most recent and useful innovations and emphasises applications to practical problems.

The emphasis in each chapter has been placed on the transducer aspect, i.e. on the instrumentation necessary to perform specific tasks, so that all the necessary components—basic theory, practical details and devices, application to actual problems—are included, as well as information concerning probable sensitivity, accuracy, etc. Simple explanations of complex physical phenomena have been used instead of rigorous treatments, the latter usually being available from the references associated with each chapter.

Engineers and applied scientists are often faced with the measurement of a wide range of parameters, e.g. dimension, displacement, strain, force, pressure, torque, fluid flow, fluid level, time dependent effects, etc., and optical methods may seem inappropriate at first glance, but all those mentioned are capable of evaluation using optics and most physical parameters are susceptible to this type of measurement.

The main advantages of these methods can be summarised as follows: no physical contact; large field coverage; very high sensitivity; applications in hostile environments. It is often one or more of these reasons which justify the use of optics.

Finally, I express my sincere thanks to all the authors who have contributed to this book.

A. R. LUXMOORE

CONTENTS

Contents

LIST OF CONTRIBUTORS

S. J. BENNETT

Division of Mechanical and Optical Metrology, National Physical Laboratory, Teddington, Middlesex TW11 0LW, UK

R. G. W. BROWN

Royal Signals and Radar Establishment, St. Andrews Road, Great Malvern, Worcestershire WR14 3PS, UK

J. N. BUTTERS

Department of Mechanical Engineering, Loughborough University of Technology, Leicestershire LE11 3TU, UK

P. W. HARRISON

3 Nightingale Road, Hampton, Middlesex TW12 3HU, UK

W. KING

Department of Mechanics of Materials, University of Strathclyde, James Weir Building, 75 Montrose Street, Glasgow G1 1XJ, UK

A. R. LUXMOORE

Department of Civil Engineering, University College, Applied Science Building, Singleton Park, Swansea SA2 8PP, UK

D. R. MOORE

AVT Engineering Services, 27 Bramhall Lane South, Bramhall, Stockport, Cheshire SK7 2DN, UK

E. R. PIKE

Royal Signals and Radar Establishment, St. Andrews Road, Great Malvern, Worcestershire WR14 3PS, UK

E. R. ROBERTSON

Department of Mechanics of Materials, University of Strathclyde, James Weir Building, 75 Montrose Street, Glasgow G1 1XJ, UK

A. T. SHEPHERD

Ferranti Ltd, Thornybank Industrial Estate, Dalkeith, Lothian EH22 2NG, UK

J. WATSON

Department of Electrical and Electronic Engineering, University College, Applied Science Building, Singleton Park, Swansea SA2 8PP, UK

Chapter 1

PHOTODETECTORS AND ELECTRONICS

J. Watson

Department of Electrical and Electronic Engineering,
University College, Swansea, UK

1. INTRODUCTION

A photosensor and its associated electronics for use in fringe counting must fulfil several functions, viz.,

(i)   it must be capable of recognising the presence or absence of a fringe; that is, it must be able to distinguish between the light and dark (i.e. ambient) regions produced by the associated optical system;

(ii)   It must also operate down to d.c. in cases where the fringes can form a stationary pattern (and where chopper, or a.c. carrier, techniques are not used); and

(iii) It must be sufficiently fast in operation to make possible the counting of fringes which appear at the highest frequency dictated by the system involved.

These requirements are not as simple as might appear at first sight, and a choice of sensor and associated electronics should be carefully made with due regard to the system as a whole. For example, requirement (i) implies that when a bright fringe illuminates a sensor with a luminous incidence, $E_{v(1)}$, then that sensor must be capable of producing a signal of a magnitude above that required to trigger the electronics, so producing a count. On the other hand, in the presence of a 'dark' fringe, which will actually illuminate the sensor at some lower level, $E_{v(d)}$, then the sensor output must be low enough not to trigger the electronics. This apparently self-evident situation is complicated by the existence of electrical noise, for if the sensor output plus noise, when illuminated at the lower level, $E_{v(d)}$, can instantaneously approach $E_{v(1)}$, then a spurious trigger pulse might occur. Also, a possible, but less probable result might be that a noise peak in the opposite sense to the signal at $E_{v(1)}$ could conceivably trigger the electronics momentarily into the OFF state, so that the return to ON would count as a second pulse.

This leads to the question of the noise inherent in the detector in question, which will be considered later. It also leads to the question of noise immunity, and this is best illustrated as in Fig. 1. Here, it is assumed that the sensor output can be represented by a signal level within a spread of noise. Starting at the left-hand side of the diagram, it is apparent that as the illumination level falls, the output band falls with it, and eventually crosses the lower trigger level of the electronic system. This means that the output band must then rise to the upper trigger level before a count is registered, so that for slow counting, the noise peaks will not produce spurious trigger pulses. That is, the provision of a hysteresis circuit greatly improves the spurious count situation (see below).

If the optical parameters are such that the noise is negligible compared with the signal, then the upper and lower trigger levels could occupy the same position, which implies the use of a simple crossover circuit.

Requirement (ii) is in part covered by the foregoing paragraph, the implication being that when an immobile pattern occurs, spurious counts or resets should not occur. That is, the level detector electronics should involve a d.c. coupled circuit, which is in fact normal in a simple level detector with or without hysteresis.

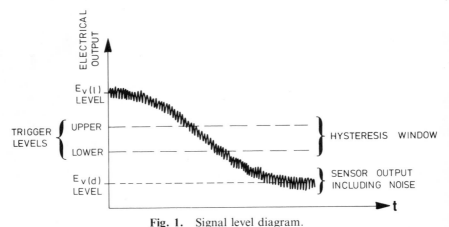

Fig. 1.   Signal level diagram.

Finally, both the detector and the following electronics must be capable of responding to the highest frequency dictated by the optical system, according to requirement (iii). That is, when the fringes are moving at their maximal velocity, both the sensor and the electronics must be capable of recovering after each count so as to record the next count. This implies that the bandwidth of the complete system must be greater than the maximum expected rate of crossing of the fringes across the detector aperture, otherwise lost counts will occur.

It is not normally difficult to acquire or design electronics to accommodate the foregoing requirements; and in recent years, the choice of sensor has narrowed as small, sensitive and fast solid-state photo-detectors have become available.

2. PHOTOSENSORS

All photosensors suitable for fringe detection fall into two categories; those which depend on the external and the internal photoelectric effects.

The external photoelectric effect refers to the phenomenon whereby a photon having an energy greater than the work function, W_f, of a given material may remove an electron from the surface of that material; that is, when,

$$hv \geq W_f \tag{1}$$

where the photon energy is given by the product of Planck's constant, h and the frequency, v.

This mechanism is involved in vacuum- and gas-filled phototubes, where the electrons emerging from a cathode are accelerated towards, and collected by, an anode. In the case of gas-filled tubes, electron-ion pairs are also produced as the primary electrons collide with atoms of the gas, so that electron multiplication occurs. A better method of electron multiplication is achieved by the photomultiplier tube, where the primary electrons are accelerated towards a positive dynode, with which they collide, so producing showers of secondary electrons. This process is repeated over a number of dynodes, each of which is at a higher positive voltage than the preceding one, so that by the time the final batch of electrons is collected by the anode, a multiplication of many orders of magnitude has been achieved.

The photomultiplier is a fast and efficient detector, and ideally suited to fringe counting in principle. However, it is also large, and requires a high voltage power supply, which contribute to somewhat problematical mechanical design and to cost. For these reasons, the use of solid-state sensors has become common, and these rely upon the internal photoelectric effect.

Here, an incident photon must have sufficient energy to raise an electron from the valence band to the first conduction band in a semiconductor material:

$$hv \geq W_g \qquad (2)$$

The band-gap energy, W_g, is generally less than the work function, W_f, even for low work-function materials such as caesium. This means that whereas the external photoeffect makes possible phototubes primarily sensitive to the energetic photons of the UV and blue regions, the internal photoeffect makes possible solid-state detectors sensitive over the red and IR regions of the spectrum. For example, W_g for silicon is about 1·1 eV, which implies that photons having wavelengths shorter than 11 180 Å could effect electronic transitions. On the other hand, caesium has a work function of about 1·9 eV, showing that photons having wavelengths shorter than 6450 Å would be needed to effect electron emission.

The foregoing figures, though illustrative of the general situation, should not be taken as limiting, for much can be done to modify both the work function of a material and also the band-gap. In the latter case, the *intrinsic* properties of the silicon (or any other semiconductor) can be markedly changed by *doping*, or including atoms of appropriate foreign materials. This results in *extrinsic* semiconductor properties, and the spectral response curve of the material can be tailored to suit numerous applications.

Insofar as fringe counting is concerned, a response within the visible region is usual, and this is easily provided by both of the two basic solid-state photosensors, the *photoresistors* and the *photodiodes*.

3. PHOTORESISTORS

This form of sensor consists essentially of a film of material whose resistance is a function of incident illumination. This film can be either intrinsic or extrinsic semiconductor material, but for the visible part of the spectrum, the chalcogenides, cadmium sulphide and cadmium selenide, are the most common.

The size and shape of the active film determines both the dark resistance and the sensitivity, and a very large range of configurations is currently available. However, for fringe counting, a small strip of material upon which the fringes can be focused is most suitable, and this strip must be connected to a pair of electrodes.

The resistance, R_S, of the cell decreases as the illumination is increased, so that it is common to consider the inverse of this resistance or the *conductance*, G_s, as being the basic parameter, which accounts for the alternative name of the sensor, the *photoconductive cell*. Electrically, therefore, it is resistance or conductance changes which must be detected by the following electronics. This implies that a current must be passed through the sensor, and either variations in this, or in associated voltage drops, should be measured.

The actual methods of making such measurements are many and varied, and two are illustrated in Figs 2(a) and (b).

In Fig. 2(a) a voltage source, E, results in sensor current, I_s, where

$$I_s = \frac{E}{R_S + R_L}$$

and R_L is a load resistor.

The output voltage across the load resistor is V_{out} where

$$V_{out} = \frac{ER_L}{R_S + R_L} \tag{3}$$

This may be measured using an amplifier with an input resistance high in comparison with R_L or the dark value of R_S.

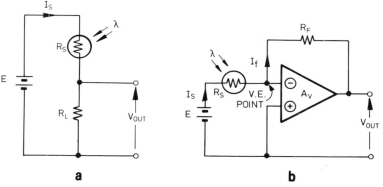

Fig. 2.   Two basic photoresistor circuits.

Equation (3) shows that V_{out} is actually an awkward function of R_S, but the circuit is nevertheless satisfactory for fringe counting, because V_{out} will be high for the high illumination condition, and low for the low illumination condition. Hence, it can be very conveniently followed by a hysteresis comparator as previously suggested.

An alternative method, which actually measures the cell conductivity, is shown in Fig. 2(b). Here, the cell current flows to the virtual earth point (or summing point) of an operational amplifier, and because this point never departs from the voltage of the common line, then the full source voltage, E, always appears across the photoresistor. Also, if the amplifier has a high input resistance, the magnitude of the feedback current, I_f, must always be equal to the magnitude of the cell current, I_s. These facts combine to give the output voltage,

$$V_{out} = I_f R_F$$
$$= -I_s R_F$$

but

$$I_s = E/R_S$$

so

$$V_{out} = -\frac{E R_F}{R_S}$$
$$= -E R_F G_s \tag{4}$$

That is, V_{out} is proportional to the conductance, G_s, so that if G_s is reasonably proportional to the incident illumination, then so is the output voltage.

It is also possible to incorporate a photoresistor into a Wheatstone Bridge circuit, but this is rarely necessary, because it is particularly suited to the detection of very small changes in resistance which would not normally be encountered in fringe counting applications.

The use of photoresistors in fringe counting is in fact very limited, because (a) they are inherently slow devices and (b) their sensitivity is somewhat dependent upon their past history and is therefore not entirely consistent. In the case of CdS cells, the relevant time constants are of the order of 100 ms, and about 10 ms for CdSe cells. Here, 'time constant' has been purposely left undefined on the grounds that not only do different manufacturers define it in different ways, but it is dependent upon applied parameters such as the actual light levels used. However, the crude figures

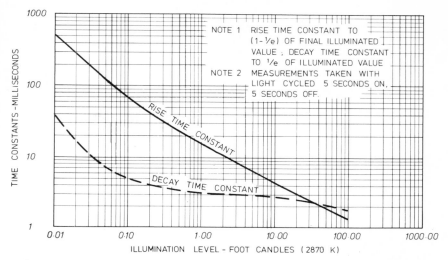

Fig. 3. Rise and decay time constant curves for the NSL-367 CdSe photoresistor. (Reproduced by permission of Silonex Inc., Montreal.[1])

quoted indicate that it would be unwise to use photoresistors to count at rates greater than a few hundred pulses per second, and even then, purpose-built photochopper cells should be specified (see Fig. 3).[1]

4. PHOTOJUNCTION SENSORS

If a single crystal of a semiconductor such as silicon or germanium is doped with both donor and acceptor impurities to form P and N regions, then the junction between these regions causes the crystal to exhibit diode properties. In the present context, it is not pointful to enter into an explanation of these properties, but only to examine their consequences.

Figure 4(a) shows the symbol for a photodiode, and it serves to define the direction of forward current (in the direction of the arrow in the diode symbol) and hence the sense of the forward voltage drop. Figure 4(b) shows a family of photodiode characteristics which indicates that operation in any of three quadrants is possible.

As an approximation, the diode dark characteristic may be represented as follows:

$$I = I_0 \left[\exp \left(\frac{qV}{kT} \right) - 1 \right] \qquad (5)$$

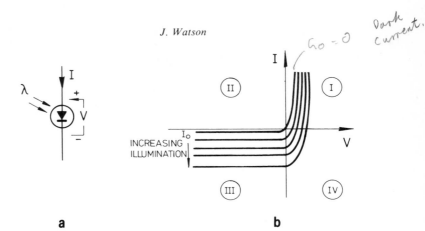

Fig. 4. The photodiode and its characteristics.

where I_0 is the reverse leakage current, q is the charge on an electron, k is Boltzmann's constant and T is the absolute temperature (K). (At a room temperature of 25 °C, kT/q is about 26 mV.)

In the first quadrant, if the forward voltage, V, is more than about 100 mV, then $qV/kT \gg 1$ and I rises rapidly and exponentially with V. Hence, the forward current of a diode can be large, and limited only by the external circuiting.

In the third quadrant, where V is negative, $qV/kT \ll 1$, so $I = I_0$. This reverse, or leakage current can be very small (down to the picaamp region), so that the diode, in effect, blocks reverse currents.

However, if energy in the form of either heat or light is applied to the diode, then I_0 increases and becomes almost a linear function of the irradiation received. This accounts for the almost horizontal family of curves seen in the third quadrant. Thus, the diode can be used to measure temperature or illumination.

As a photodiode, the device is said to be operating in the *photojunction* mode when a reverse voltage is applied, and it is responding to illumination. Under these circumstances, its internal resistance is very high, as evidenced by the nearly horizontal family of reverse current characteristics.

This reverse current is now called the *photocurrent*, I_p, and the photodiode can be represented by a perfect diode in parallel with a current generator as shown in Fig. 5(a). If the simple circuit of Fig. 2(a) is now employed, as in Fig. 5(b), then because the photodiode internal resistance is so much larger than R_L, then

$$V_{out} = (I_0 + I_p)R_L \qquad (6(a))$$

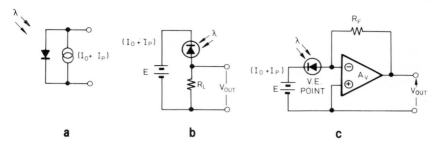

Fig. 5. The photodiode equivalent (a) and two reverse bias photojunction mode operating circuits (b) and (c).

and at levels of illumination such that $I_p \gg I_0$ (or if $I_0 \to 0$), then this becomes

$$V_{out} = I_p R_L \qquad (6(b))$$

If V_{out} is measured with a very high input resistance circuit, then V_{out} is proportional to the illumination level. However, it is better to measure $(I_0 + I_p)$ directly, as in Fig. 5(c). Here,

$$V_{out} = -(I_0 + I_p)R_F \qquad (7(a))$$

or if $I_p \gg I_0$ (or $I_0 \to 0$),

$$V_{out} = -I_p R_F \qquad (7(b))$$

In this circuit, the photodiode is working into a virtual earth point, that is, into a very low resistance given by R_F/A_v where A_v is the forward voltage gain of the operational amplifier used.

Returning to Fig. 4, it will be seen that the photodiode characteristics also exist in the fourth quadrant. In particular, it will be noted that a photocurrent can flow even when the applied voltage is zero. This is because at a P–N junction, an internal potential difference is generated so that charge separation still occurs. Working into a low resistance circuit this self-generated photocurrent is reasonably linear with illumination. However, if the load resistance is very high (approaching an open circuit), then the *photovoltage* which appears is roughly logarithmic with illumination.[2]

Working in the ways described above, and with no external voltage source, the photodiode is said to be a *photovoltaic cell*.

In the photojunction mode (third quadrant), the applied voltage is usually of the orders of volts or tens of volts, and so is much larger than the few hundreds of millivolts typically generated in the photovoltaic mode. This is why Fig. 4(b) is actually not to scale, and it is common practice to show each quadrant separately, with each drawn as if it were in the first quadrant. Figure 6 shows how this is done, and each family of characteristics is shown alongside a photodiode symbol and circuit which defines the conventional voltage and current directions. The real directions are then indicated by the signs of the sample values of V and I shown on the graphs.

In Fig. 6(a), the concept of *load-line* is also illustrated. Here, the applied reverse voltage is E (shown as -5 V in this case), and with reference to Fig. 5(b), the slope of the load-line is R_L (2500 Ω in this case). The operating point of the photodiode is that at which the characteristic relevant to the prevailing level of illumination crosses the load-line.

Note that if $R_L \to 0$, as is the case when the photodiode works into a virtual earth point as in Fig. 5(c), then the load-line would be vertical.

Figure 6(b) shows the fourth quadrant photovoltaic characteristics transposed to the first quadrant. In this mode of operation, the basic circuits of Figs 5(b) and (c) may again be used, *but with no applied voltage.* If R_L is made high in the Fig. 5(a) method, then the load-line lies near the voltage axis of Fig. 6(b) as shown, and the output voltage is simply the cell voltage, which increases logarithmically with illumination. Conversely, if

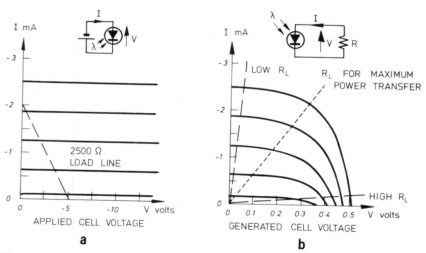

Fig. 6. Photojunction (a) and photovoltaic (b) characteristics plotted in first quadrant. (Conventional voltage and current directions shown inset.)

the photovoltaic cell works into a near-zero load resistance (like the virtual earth point in Fig. 5(c)), then the load-line is almost vertical, and the generated current varies linearly with illumination.

It is of interest to note that there will be a load resistance value (shown in Fig. 6(b)), at which maximum electrical power is transferred to R_L. This occurs when R_L has the same value as the slope resistance of the photodiode, and represents the optimum condition for its operation as a solar cell.

5. PHOTODIODE CONSTRUCTION

When P- and N-doped regions in a semiconductor crystal meet, a *depletion layer* is formed. This means that because of the physical processes occurring in the doped crystal, a separation of negative and positive charges (i.e. electrons and holes) takes place. Hence, an electric field gradient is set up, and the junction region becomes depleted of the mobile charges represented by free electrons or holes, because they are swept to one side or the other along the electric field gradient. Actually, for such a diode, equilibrium conditions are set up in which contact potentials are countered by the internal field, so that no external voltage can be measured. However, if the diode is illuminated, then given the conditions of eqn (2), hole–electron pairs can be formed by incoming photons, and a current or voltage can then be measured in an external circuit. This is what constitutes photovoltaic operation.

Practically all of the electric field gradient exists in the depletion layer, so that it is very largely here that charge separation occurs. Should the incoming photons release hole–electron pairs in either the P region or the N region, the chances are that only recombination will take place, which cannot contribute to the current. Hence, it is desirable to increase the thickness of the depletion layer, and one way of doing this is to apply a reverse voltage which augments the internal field. This results in photojunction operation, as has been described.

Incoming photons penetrate to depths dictated by their energies, the longer wavelength photons penetrating furthest. So, if the P region is encountered first, it should be thin, making for a good response at the blue end of the spectrum. This suggests the use of a thin layer of low-resistivity (i.e. heavily-doped) material. However, full utilisation of the more penetrating lower energy photons means that a thick depletion layer should follow, which implies a high resistivity (i.e. lightly-doped) material. Finally

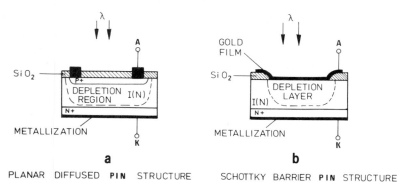

a   **b**

PLANAR DIFFUSED **PIN** STRUCTURE   SCHOTTKY BARRIER **PIN** STRUCTURE

Fig. 7.   Idealised cross-sectional representations of PIN photodiodes (not to
scale).

a low resistivity N-type material should follow for good bonding to the
electrode metallisation. These conflicting requirements can be met by
sandwiching a layer of high resistivity, near-intrinsic or I-type material
between low resistivity P^+ and N^+ layers to form a PIN (or NIP)
photodiode. A planar diffused PIN photodiode structure is shown in
idealised form in Fig. 7(a). Here, the light enters the thin P^+ region via the
transparent SiO_2 passivating layer, and is absorbed largely in the I-type
region. The anode is formed as a metal annulus over the P^+ region and the
cathode as a film deposited over the N^+ region.

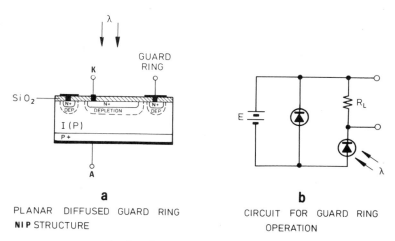

a   **b**

PLANAR DIFFUSED GUARD RING   CIRCUIT FOR GUARD RING
NIP STRUCTURE   OPERATION

Fig. 8.   The guard-ring photodiode.

For even better response at the blue end of the spectrum, the thin P^+ region can be omitted, and a transparent gold film deposited directly on to the near-intrinsic N region. The metal–semiconductor junction is known as a Schottky barrier, and also results in a depletion layer being formed, as shown in Fig. 7(b). Further, the gold film acts also as the anode.

In all semiconductor devices, unwanted surface currents occur, and though small, these can be further reduced by the use of a *guard ring* as shown incorporated into an NIP structure in Fig. 8(a). Here, not only is a circular N^+ region diffused into the near-intrinsic P material but so is an N^+ annulus. Both of these heavily-doped N regions are brought out to electrodes, and Fig. 8(b) shows how they are connected. Any surface current is collected by the guard ring and fed directly to the voltage source, whilst the current collected by the normal active N^+ region must pass through the load resistor. This type of photodiode is used where low dark or leakage current is mandatory.

6. PHOTODIODE CHARACTERISATION

There are numerous parameters which serve to characterise photodiodes, the most important of which are:

(i) physical dimensions (notably active area);
(ii) mode(s) of operation;
(iii) sensitivity (e.g. responsivity);
(v) speed of operation;
(vi) noise characteristics (e.g. noise equivalent power (*NEP*) and specific detectivity).

Taking these in order, obviously the physical dimensions will be dictated by the system. Usually, photodiodes have circular active areas (by contrast with photoresistors), but there is also a range of position-sensitive devices and some of these will be considered later.

The modes of operation for a photodiode are simply either photo-junction operation with a reverse biasing voltage, or photovoltaic operation with zero biasing. Because reverse biasing implies a thick depletion region with a high electric field gradient, it makes for a faster response. It also ensures that the longer wavelength photons generate their electron–hole pairs in the depletion region, so that the spectral response is improved. Also, by virtue of operating the diode well into the third quadrant the region where the characteristics bend over into the photovoltaic region is avoided, making for good linearity.

Conversely, the photovoltaic mode avoids the need for a voltage source, and provides good low-noise operation up to about 100 kHz (for a small-area device with consequent low capacitance), and the loss in linearity does not usually matter if only the presence or absence of a fringe is to be detected. For faster operation, or when a good waveshape has to be obtained from a chopped light beam, the reverse-biased photojunction mode is preferable.

It should be noted that in much of the literature, 'photoconductive' operation refers to the reverse-biased operation of a photodiode. This is an unfortunate term, because the relevant photodiode characteristics are those of a current generator, not a conductor. In this chapter, the term 'photoconductive' is used only in relation to the photoresistor, which is a genuine photoconductive cell having ohmic properties.

The sensitivity of a photodiode can be expressed in several different ways, of which the most fundamental is the *quantum efficiency*, η. This is defined as the average number of electrons released per photon, which is obviously less than unity because not all the photons which reach the cell produce hole–electron pairs, and some of the pairs are lost by recombination.

However, the number of photons reaching the cell is not a commonly used quantity: it is more usual to represent the incoming radiation in terms of either radiant incidence (irradiance), or luminous incidence (illuminance). The first term, irradiance, E_e, refers to the energy (at all wavelengths) per unit time per unit area which arrives from a direction normal to the active surface. The SI units are therefore watts per square millimeter.

Illuminance, E_v, is defined in the same way except that only the visible part of the spectrum is involved. This implies that *spectral* irradiance must be weighted according to the spectral sensitivity of the human eye, to give the illuminance.

Sensitivity can now be defined in terms of the responsivity, which is the photocurrent produced (under specified working conditions) per unit radiant or luminous flux for a cell of effective area, A:

$$R_{\phi e} = \frac{I_p}{E_e \cdot A} \quad \text{(amps per watt)} \quad \text{(8(a))}$$

or

$$R_{\phi v} = \frac{I_p}{E_v \cdot A} \quad \text{(amps per lumen)} \quad \text{(8(b))}$$

Here, in both cases, the total power falling on the active area of the cell is given by multiplying power density, E_e or E_v, by the effective cell area, A. In the latter case, the illuminance, E_v, is in lumens per unit area, where the lumen is the unit of light power or luminous flux. That is, it is the radiant power in watts weighted according to the sensitivity of the human eye as explained above. At the wavelength of peak sensitivity of the light-adapted human eye (5550 Å), 1 W = 680 lumens.

If the incoming irradiation is derived from a source of narrow bandwidth such as a monochromator, then the sensitivity can be plotted at a series of wavelengths, resulting in a spectral sensitivity curve.

Typical spectral sensitivity curves for silicon PIN structures are given in Fig. 9, along with a plot of the spectral sensitivity of the average human eye for comparison purposes.

If a photodiode is used in the reverse-biased (photojunction) mode, then it is linear over many decades of irradiance, and departures from linearity are more likely to result from the dynamic range limitations of the associated electronics rather than from the diode itself. Switching methods (e.g. changing the value of R_f in Fig. 5(c)) can effectively increase this dynamic range; alternatively, photovoltaic operation into a very high resistance will result in a logarithmic response with consequent range compression.

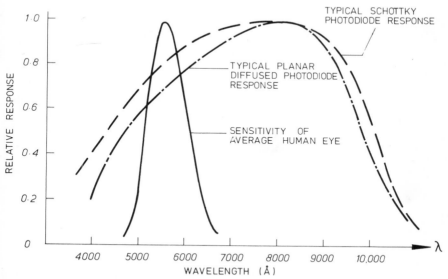

Fig. 9. Some typical spectral sensitivity curves.

Insofar as speed is concerned, photovoltaic operation is comparatively slow because the depletion layer is thin, as mentioned previously. However, for reverse bias photojunction operation, response times well below a microsecond are obtainable. The definition of response time and how it is measured varies between manufacturers, but small area fast photodiodes have response times which are limited largely by the external circuitry and its stray capacitances.

Finally, the question of noise arises, which is inevitably bound up with sensitivity because the lower limit of sensitivity must be defined by the level of irradiance which gives rise to a photocurrent similar in magnitude to the noise level. The noise itself can best be explained by reference to an electrical equivalent circuit for the photodiode.

7. THE EQUIVALENT CIRCUIT AND NOISE CHARACTERISATION

Figure 10 shows an equivalent circuit for a guard-ring photodiode,[3] the components of which are:

(a)   a current generator representing the photocurrent, I_p;
(b)   a current generator representing the leakage current, I_0;
(c)   a dynamic or slope resistance, r_d, representing the slope of the characteristic (Figs 4 or 6) at the relevant working point;
(d)   a capacitance, C_j, representing the capacity of the junction;
(e)   the bulk series resistance of the leads and the photocell material, R_{sb};
(f)   the channel resistance between the active and the guard-ring diffusions, R_{ch} (which will be absent for a photodiode having no guard-ring).

The combination of C_j, R_{sb}, R_{ch} and the external load resistance and capacitance, will define the speed of response. The value of C_j is least at high reverse voltages, hence the superiority of photojunction operation in this context.

Under dark conditions (particularly in the photojunction mode) the leakage current I_0 will give rise to shot noise according to the Schottky equation:

$$\overline{i_{no}^2} = 2qI_0\,\Delta f \tag{9}$$

where $\overline{i_{no}^2}$ = mean square shot noise, q = electronic charge, Δf = noise bandwidth.

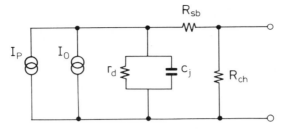

Fig. 10. An equivalent circuit for a guard-ring photodiode.

In the equivalent circuit, both R_{sb} and R_{ch} contribute thermal or Johnson noise, but this is usually dominated by the shot noise (r_d contributes no noise, being simply a resistance slope, and nor does the capacitance, C_j).

The shot noise can now be represented in terms of unit noise bandwidth, or *spot noise*:

$$\bar{i}_{n0} = \sqrt{(2qI_0)} \tag{10}$$

typical units being nanoamps per root hertz ($nA/\sqrt{Hz}$).

At this point, it should be noted that at operating frequencies below a few hundred hertz, *flicker noise* appears. This is sometimes known as *excess noise* or $1/f$ *noise*, because it exceeds the shot noise at low frequencies, and because it increases inversely (very approximately) with frequency. At very low frequencies, approaching d.c., this noise may be regarded as drift relative to a d.c. component, the reverse leakage current itself.

Returning to the shot noise region, the noise equivalent power (*NEP*) is defined as that level of irradiance which would produce a signal current I_p equal to the magnitude of the noise current. If the noise current is taken as simply $\bar{i}_{n0}$, then from eqn 8(a), the incoming power, or radiant flux for a signal of this level would be the *NEP*:

$$NEP = \frac{\bar{i}_{n0}}{R_{\phi e}} \tag{11}$$

Note that this expression refers to the unit noise bandwidth (or spot) condition, so the units of *NEP* will be in watts per root hertz, or more practically, $fW/\sqrt{Hz}$ where fW stands for femtowatts, or watts $\times 10^{-15}$.

If the cell were now irradiated at some level $E_e . A$, then the spot signal-to-noise ratio would now be:

$$\left(\frac{S}{N}\right) = \frac{E_e . A}{(NEP)} \tag{12}$$

In this expression, the signal increases with the effective area of the photodiode. Also, because the leakage current also increases with the area, then so does the mean square of the noise current, accordingly to the Schottky eqn (9). Hence, the noise current itself $\overline{i_{no}}$ increases with the root of the area, and therefore so does the NEP. So, in eqn (12), the S/N ratio must also increase with the root of the area.

The *specific detectivity*, D^* (pronounced 'dee-star'), is defined by normalising the S/N ratio with respect to area, and to irradiance, by dividing by $\sqrt{A}$ and by E_e. This gives, using eqn (12),

$$D^* = \frac{(S/N)}{\sqrt{A}.E_e} = \frac{\sqrt{A}}{NEP} \qquad (13)$$

and is now independent of any particular cell size (or type). A practical unit would be $cm\sqrt{Hz}/W$.

The operating conditions under which both NEP and D^* are quoted must always be given, and the convention is that bracketed numbers follow the magnitudes of these quantities. These numbers are:

(a) the wavelength of the irradiance;
(b) the frequency at which the measurement is made (i.e. chopper frequency);
(c) the noise bandwidth, Δf.

For example, E, G & G[4] photodiode type SGD-040 has a quoted NEP of:

$$NEP = 9.6 \times 10^{-14} \, W \qquad (0.9 \, \mu m, 10^3, 1)$$

which means that this NEP was measured using light of 9000 Å (i.e. in the near infrared), with a chopper frequency of 1000 Hz, and has been converted to a spot value or unity noise bandwidth.

The active area of this device is quoted as $0.82 \, mm^2$, so that the D^* can be obtained using eqn (13). Converting the area to $0.0082 \, cm^2$, this is:

$$D^* = \frac{\sqrt{0.0082}}{9.6 \times 10^{-14}} \simeq 0.0094 \times 10^{14}$$

Expressed fully, this would be,

$$D^* = 9.4 \times 10^{11} \, cm\sqrt{Hz}/W \qquad (0.9 \, \mu m, 10^3, 1)$$

8. POSITION-SENSITIVE PHOTOSENSORS

Although several methods have been employed in the realisation of position-sensitive cells, two comparatively simple analogue forms are commercially predominant, in addition to the complex multistructure and charge-coupled device (c.c.d.) types now becoming generally available.

The most common position-sensitive cell (PSC) is the split photodiode, which in its simplest form consists of a normal photodiode whose active area is split into two parts, so that it is effectively two photodiodes in close juxtaposition, and with one common electrode. Figure 11(a) is an idealised representation of such a sensor, whilst Fig. 11(b) illustrates its usage in the photojunction mode.[3]

Here, the two photocurrents have been passed through identical load resistors, in order that the salient features of the system might be easily demonstrated.

The first point to note is that outputs V_A and V_B exist concurrently only when an incident light spot falls across the split and energises both diode halves. When the light impinges entirely on one side, then no further positional information is available. That is, positional information is available only over a distance equal to twice the diameter of the light spot.

Secondly, the obvious way to obtain a position-dependent output is to

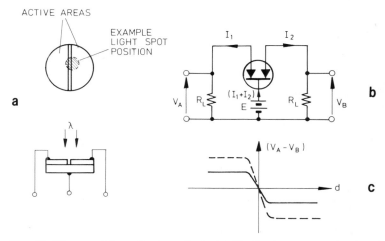

Fig. 11.   Position-sensitive split photodiode. (a) Idealised representation; (b) dual photojunction connection; (c) difference output ($V_A - V_B$) versus distance from split, d.

plot the difference of the two individual outputs ($V_A - V_B$), and this is done in Fig. 11(c) which clearly shows how the cell operates simply as two separate diodes when the spot has moved off the split.

Two disadvantages of the method now become apparent. Firstly, the linearity of the output as a function of spot position is dependent upon the shape of the spot and homogeneity of the illuminance which it produces; and secondly, the steepness of the curve is proportional to the average illuminance. A dotted curve, representing the response to a brighter spot, is shown in Fig. 11(c) to illustrate this latter point, and later it will be explained how an appropriate electronic circuit can overcome this problem.

Fabrication methods for large area photodiodes are sufficiently flexible to allow the manufacture of numerous other configurations. For example, multielement one-dimensional arrays are available,[5] as are four-quadrant (or two-dimensional) sensors, in addition to matrix and almost any conceivable customised forms.

Some of the disadvantages of the split photodiode—notably that of limited positional range—are overcome to a considerable extent by the Posicon.* This consists essentially of a Schottky barrier structure[6] having the (idealised) cross section shown in Fig. 12(a). When the incoming light penetrates the transparent gold film, it releases hole–electron pairs in the intrinsic region. The resulting current is divided between the two edge-mounted ohmic contacts according to the resistances of the bulk material between each of these electrodes and the mean carrier generation point. Hence, the difference between the two currents is proportional to the light spot position over a wide range, and the resulting characteristic is as in Fig. 11(b).

Because the light spot does not have to traverse a discontinuity, its shape does not matter a great deal; and output variations due to changes in intensity can be electronically minimised as for the split photodiode.

Insofar as specifications of linearity and positional sensitivity are concerned, the nature of the associated optical systems must first be closely defined, otherwise any such figures will be meaningless irrespective of the type of sensor involved. For example, the linearity of the Posicon* is quoted as 1 % over the middle 10 % of the device length, and the user should refer to the manufacturer for the conditions under which this is valid.

As has been mentioned, the split photodiode is available in four-quadrant form for two-dimensional position-sensing, and its properties

* Posicon is a trademark of United Detector Technology Inc.

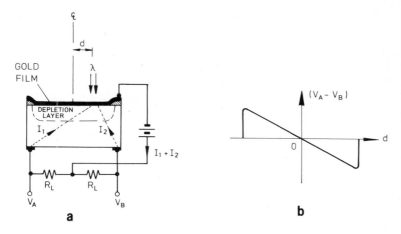

Fig. 12. Principle (a) and output characteristic (b) of the United Detector Technology 'Posicon' position-sensitive cell.

under these circumstances are easily extrapolated from the foregoing presentation. The Posicon is available with four edge electrodes, and this also makes possible two-dimensional sensing. It is useful to compare the performance of this latter device with that of the Wallmark sensor or lateral diode which consists essentially of a semiconductor chip having four ohmic contacts, plus a central P–N junction on one side. The light spot is focused on to the other side, and relative outputs from the four ohmic contacts define its position.[7]

The devices described above are simple and easy to use, and will fulfil most position sensing requirements. However, another mode of photo-diode operation makes possible more sophisticated positional measure-ments, including those where only low light levels are available. This is the charge storage mode, and it has led to the availability of both linear and matrix arrays of several thousand photodiode elements fabricated by integrated circuit techniques.[8]

Briefly, if a photodiode is illuminated, it will charge up its own capacitance to a voltage dictated in part by the level of illumination and in part by the inherent leakage current. Hence, if a series of such diodes is sequentially interrogated, the voltage levels can be extracted as a series of pulses of heights dependent upon the incident illumination.

Such devices can form the basis of pattern recognition systems and, in principle at least, could lead to image transduction. However, the latter possibility has been realised using charge-coupled arrays, in which packets of photogenerated charges are moved through a homogeneous (single crystal) substrate, as opposed to a matrix of individual photocell structures.

Because these techniques are somewhat specialised in the present context however, further details will not be presented, and the interested reader is referred to ref. 9 at the end of this chapter.

9.   BASIC PHOTOSENSOR ELECTRONICS

Because the output voltage or current of a photosensor is small with levels of irradiation normally encountered, it is usual to follow such a transducer with some form of amplifier. Where irradiation measurement or position sensing is concerned, for example, an analogue output is required, so that the amplifier must be capable of accepting all signal levels within the expected dynamic range, and amplifying them with minimal distortion. On the other hand, in applications such as fringe counting, the amplifier may be designed to saturate in the presence of a fringe, so that an essentially digital output is produced, which may then be applied to a commercial pulse counter.

Consider first the analogue operation of a photodiode and its associated amplifier.[10]

For zero external biasing, and with a low value of load resistance (corresponding to the 'low R_L' load-line in Fig. 6(b)), the self-generated output current of the photodiode will be sensibly linear with irradiation. As has been mentioned, the input point of an inverting operational amplifier presents a very low resistance, so that the photodiode can be connected as in Fig. 13(a). The output voltage can be determined (as for eqn 7(b)) by assuming that the photocurrent, the dark or leakage current and the noise current all flow via the feedback resistor R_F as shown in Fig. 13(a):

$$V_{out} = -(I_p + I_0 + i_n)R_F \qquad (14)$$

In practice, the wideband noise current, i_n, consists largely of the shot noise due to fluctuations in the leakage current and the photocurrent. The component associated with the leakage current is small in the present case, because the lack of bias implies that this leakage current is minimal.

Note that the direction of the photocurrent is in the same direction as the leakage current, i.e. it is a reverse current, and is so shown in Fig. 13(a).

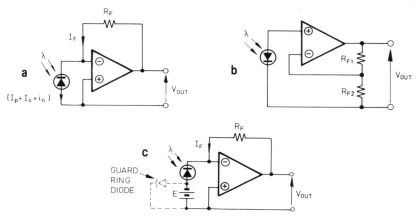

Fig. 13. Photodiode–amplifier connections. (a) Self-generating current mode; (b) photovoltaic mode; (c) reverse bias photojunction mode.

(This point is easily understood by reference to Fig. 4.) Hence, the output end of R_F will go positive so that V_{out} is also positive.

If photovoltaic operation is desired, the non-inverting configuration of Fig. 13(b) may be employed. Here, the input resistance to the non-inverting amplifier is extremely high, so that the 'high R_L' condition of Fig. 6(b) becomes valid. Under these circumstances, the response is closely logarithmic, and this can be understood by assuming that the photocurrent flows through the internal dynamic diode resistance, r_d, shown in the equivalent circuit of Fig. 10. The resulting voltage drop acts to forward bias the junction, so that the value of r_d decreases, and it does so exponentially with photocurrent and hence with irradiation. Thus, the net photovoltage is logarithmic.

The voltage gain of the non-inverting operational amplifier configuration[3] is simply:

$$A_{v(FB)} = 1 + \frac{R_{F1}}{R_{F2}} \qquad (15)$$

This is usually quite low, because the photovoltage levels are typically some tens or hundreds of millivolts.

Returning to linear operation, higher speeds can be obtained by reverse-biasing the photodiode into the photojunction region, represented in Fig. 6(a). An appropriate circuit is given in Fig. 13(c), and a guard-ring connection is shown dotted for cases where this form of sensor is used. The leakage current, and hence the noise, is somewhat higher in this mode, so that a guard ring may be advantageous when reverse bias voltages above about 50 V are employed. Equation (14) is again relevant to this mode of operation.

9.1. Digital Operation

This mode is best appreciated by taking a practical case where the voltage gain of an amplifier is so great that the output reaches its maximum and minimum possible values before the input signal reaches its own peaks. That is, *clipping* occurs. Figure 14(a) represents the output signal of an inverting, linear amplifier, whose input signal is sinusoidal. If the gain is increased to produce clipping, the trace shown in Fig. 14(b) will appear. Finally, for a very high gain, the output wave will appear like a square wave whose crossover points appear concurrently with those of the input sine wave, as shown in Fig. 14(c). This output can now be considered digital, and the amplifier is acting as a *comparator*.

An open-loop operational amplifier can be used in this way, and its output will change over whenever an input voltage crosses a reference voltage, V_{ref}, as shown in Fig. 15. (Obviously V_{ref} can be zero, in which case a *zero-crossing* comparator results.)

It will now be recognised that the preceding discussion has reverted back to the beginning of the chapter, where it was suggested that a bright fringe might be required to initiate a count. As an example, Fig. 15(c) has been included to show how a photojunction cell can be loaded by a resistor, R_L, so that when the voltage drop exceeds V_{ref}, the amplifier output will change polarity.

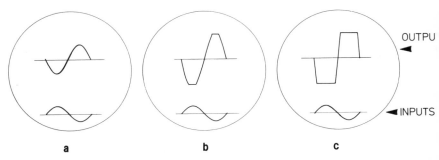

Fig. 14. Output waveforms for (a) a linear inverting amplifier; (b) the same with high gain leading to clipping; (c) the same with open loop resulting in squaring.

Here, a single-supply comparator amplifier has been chosen for simplicity, and V_{ref} will be given by:

$$V_{ref} = \frac{V_{cc}^+ R_2}{R_1 + R_2} \tag{16}$$

Note that because no shunt feedback resistor is present, then no virtual earth point is present either, so that if the internal input resistance of the comparator is high (which is usual), then R_L is the only load applied to the photodiode.

If the output voltage of the comparator can vary between almost zero and almost V_{cc}^+ then the transfer function will appear as shown in Fig. 15(d).

This simple crossover circuit with its single trigger level, V_{ref}, can be modified to provide two trigger, or reference levels, $V_{ref(hi)}$ and $V_{ref(lo)}$, so that noise immunity can be conferred, as was also explained at the beginning of the chapter.

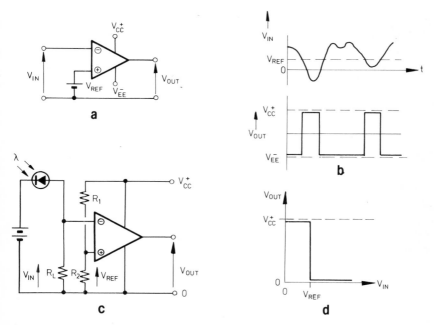

Fig. 15. The comparator or crossover detector. (a) Basic method; (b) performance; (c) photojunction application; (d) transfer function.

Figure 16(a) shows how a feedback resistor R_3 may be connected from the output to the non-inverting reference input, to realise this function. Its operation may be explained as follows.

Suppose that V_{in} is low (almost zero) so that V_{out} is high (almost V_{cc}^+). The resistor R_3 is now effectively in parallel with R_1 so that

$$V_{ref(hi)} = \frac{V_{cc}^+ R_2}{\dfrac{R_1 R_3}{R_1 + R_3} + R_2} \tag{17}$$

If now V_{in} rises above this value, then V_{out} goes low, and R_3 is effectively placed in parallel with R_2 so that,

$$V_{ref(lo)} = \frac{V_{cc}^+ \dfrac{R_2 R_3}{R_2 + R_3}}{R_1 + \dfrac{R_2 R_3}{R_2 + R_3}} \tag{18}$$

The transfer function is now as in Fig. 16(b) and a noisy input waveform would produce the output shown in Fig. 16(c), which is as required. This circuit is known as a hysteresis comparator, and the output signal may be applied to a pulse counter with little or no further processing.

9.2. Frequency Response

Having seen how basic analogue and digital outputs may be obtained, it is pertinent to consider the frequency responses of the systems, and the question of whether the photocell output will be of adequate magnitude.

In the case of analogue circuits, it is the frequency response of the amplifier which is of importance, provided of course the photosensor has an adequately rapid response. In principle, amplifiers which can accommodate signals well into the tens of megahertz are readily available, and can also be quite easily designed for individual requirements. However, in the case of operational amplifiers which will accept feedback components as described above, the situation is not quite straightforward. It can be best explained by noting that any amplifier designed to be unconditionally stable irrespective of the degree of feedback employed, must exhibit a voltage gain which varies inversely with frequency. That is, the gain-bandwidth product, E_T, must be constant. (This is because the phase-shift of such an amplifier does not exceed $90°$, so that positive feedback, which may lead to oscillation, cannot occur.)

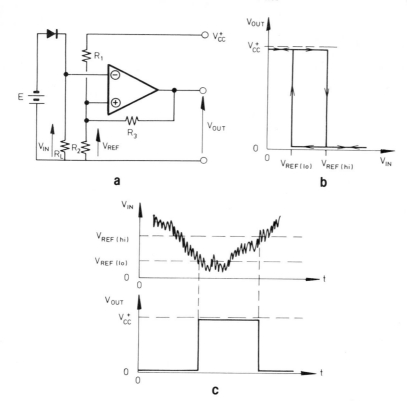

Fig. 16. The hysteresis comparator. (a) The circuit; (b) the transfer function; (c) the performance with a noisy signal.

Such a characteristic is shown in Fig. 17, from which it will be seen that the lower the gain (as defined by the feedback resistors) the wider the bandwidth or frequency range which can be accommodated. Typically, modern internally-compensated operational amplifiers have gain-bandwidth products somewhere between 1 MHz and 10 MHz, so that at moderate gains, bandwidths of a few kilohertz are easily obtainable.

In the cases of an operational amplifier used as a comparator (or a purpose-designed comparator) it is not the small-signal bandwidth which limits the frequency response, but the *slewing rate*. This defines the rate at which the output voltage can swing between its maximal values. Typically, modern integrated circuit amplifiers and comparators will be found to slew at rates from 1 V/μs to 100 V/μs, which is more than adequate for most fringe counting applications.

In cases where a light chopper is used, slewing rate is particularly important, and so is amplifier linearity. If the chopped light beam has a small cross-sectional area, the photosensor (if linear and sufficiently fast) will produce a near square wave whose height is proportional to the illuminance. Hence, the following amplifier must not degrade this square wave to a trapezoid (or triangle in the limiting case): neither must it degrade the linearity of the system significantly.

Insofar as the magnitude of a signal is concerned, it is obviously possible to calculate the expected output current of a photodiode from a knowledge of its sensitivity and the irradiance or illuminance anticipated. However, it has to be admitted that most optical systems are set up somewhat empirically in the first instance, so that the choice of photosensor may well be partly a matter of trial and error. In this context, the configuration of the light spot will determine the size and shape of the sensor, and its electrical output is then easily measured.

If the available illuminance is insufficient to produce an electrical output

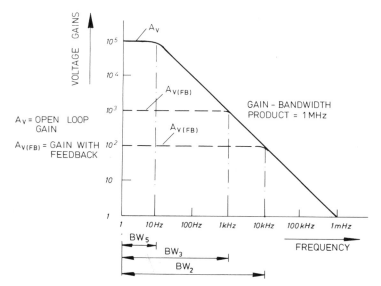

Fig. 17. Typical gain–frequency characteristic for an operational amplifier. BW_5 = bandwidth at $A_v = 10^5$; BW_3 = bandwidth at $A_{v(FB)} = 10^3$; BW_2 = bandwidth at $A_{v(FB)} = 10^2$.

large enough to drive a comparator directly, for example, then consideration should be given to specifying a hybrid photosensor containing its own preamplifier.

The simplest device in this class is the phototransistor, which consists simply of a regular transistor (which may be bipolar or field-effect)[3] in which the structure is open to incident illumination. The symbol for an NPN phototransistor, and two modes of connection, are shown in Figs 18(a) and (b). When the device is irradiated, the collector-base leakage current increases as for a simple photodiode, and this is amplified by the transistor current gain ($\bar{\beta}$ or h_{FE}) to produce a comparatively high collector current I_c. In Fig. 18(a), I_c flows through R_c, so producing a voltage drop which takes the output negative as the irradiance increases. The emitter

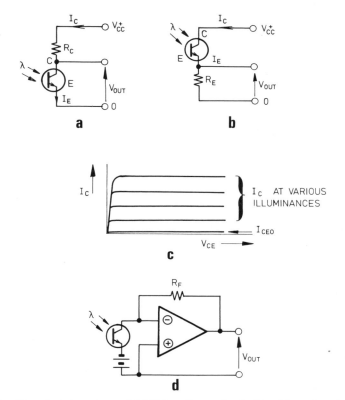

Fig. 18. The phototransistor. (a) With collector load; (b) with emitter load; (c) characteristics; (d) a speed-up technique.

current I_E is similar to I_c in magnitude so that the voltage drop across R_E in Fig. 18(b) would change in the opposite sense.

The current gain is a function of I_c, so that the phototransistor is not a linear transducer. Also, the inherent leakage current I_{CEO} is dependent on temperature, which can pose problems at low light levels where the values of I_c and I_E are comparable with I_{CEO}. It is therefore best to use the phototransistor in the saturating mode if possible.

The phototransistor also exhibits quite a poor frequency response due to the presence of internal collector-base capacitance. This may be improved by connecting the collector directly to the virtual earth point of an inverting operational amplifier,[11] as in Fig. 18(d). The collector-base capacitance is now held at a constant voltage, so that it does not have to charge and discharge with the signal, and the result is that the frequency response becomes largely a function of the operational amplifier.

A better overall photosensor results from combining a photodiode with an operational amplifier in a single encapsulation.[12] One advantage is that the short leads ensure minimal electrical interference pick-up: but in some cases, flexibility of coupling is lost because the photodiode is permanently connected to its amplifier. However, some such encapsulations do offer either analogue output, or digital output with a programmable threshold level.[13]

10. ELECTRONICS FOR POSITION-SENSITIVE CELLS

Figures 11 and 12 show that both the split photodiode and the Posicon produce two output currents which can be used to produce two voltage drops across two load resistors. Also, in both cases, it is the difference between the two currents (or their voltage drops) which must be measured to define the position of the light spot. However, in this simplistic explanation lies the implicit assumption that the *sum* of the two currents or voltages is always constant: if it were not, then for a given spot position, the positional readout would be a function of the illuminance, which is obviously undesirable. This problem can be solved in two ways. Either the ratio of the two parameters can be measured; or else the difference can be measured, and the sum electronically normalised.

Keeping in mind the previous material on methods of loading photodiodes, it is reasonable to choose current, rather than voltage measurement. Hence, the two output currents from either type of device should be fed into the (zero-impedance) virtual-earth inputs of operational

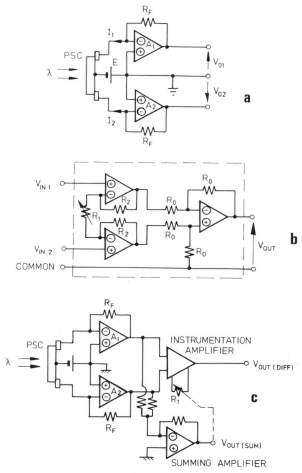

Fig. 19. Electronics for position-sensitive cells (PSCs). (a) Input transresistance amplifiers; (b) differencing instrumentation amplifier; (c) feedback system. (NB, conceptual schematics only.)

amplifiers, as shown in Fig. 19(a). By this means a current-to-voltage (or transresistance) conversion is achieved, the output voltages being $I_1 R_F$ and $I_2 R_F$, respectively.

These voltages can now be fed to an *instrumentation amplifier* which is a composite unit consisting of three operational amplifiers arranged as in Fig. 19(b). The composite unit (which is available in single-chip monolithic form) is a high input resistance differencing module whose gain is a function of the single resistance R_1 in Fig. 19(b). If the output voltages of the transresistance amplifiers are V_{01} and V_{02}, respectively, then the output voltage of the instrumentation amplifier would be:[3]

$$V_{\text{out}} = (V_{01} - V_{02})\left(1 + \frac{2R_2}{R_1}\right) \qquad (19)$$

This means that

$$V_{\text{out}} \propto (V_{01} - V_{02}) \propto (I_1 - I_2) \qquad (20)$$

which is effectively the expressions plotted in Figs 11(c) or 12(b).

As has been mentioned, the validity of the expression as a measure of position depends upon $(I_1 + I_2)$ being constant. This is an unlikely situation having regard to such factors as light-source deterioration with time and soiling of optical components, but fortunately electronic compensation can be simply achieved. If a summing amplifier (which is simply an inverting operational amplifier with multiple inputs) is used to produce an output proportional to $(I_1 + I_2)$, then its own output can be used to modify the gain of the instrumentation amplifier. Equation (19) shows that this can be achieved by controlling the value of R_1, which can be (for example) an indirectly-heated thermistor or a field-effect transistor operating as a voltage-controlled resistor. The complete system is shown in block-diagrammatic form in Fig. 19(c). Here, the transresistance and instrumentation amplifiers are shown as separate units: it is in fact possible to combine them into a single 'transresistance instrumentation amplifier' using modern multiple monolithic operational amplifiers and matched resistor networks.[14]

The method described above is but one way of achieving the desired result. Another useful concept[15] is to employ a divider module which will perform the function:

$$V_{\text{out}} \propto \frac{I_1 - I_2}{I_1 + I_2} \qquad (21)$$

Here, the summing and differencing is performed as before, and if both I_1 and I_2 change by the same factor K, then clearly K will always cancel out in eqn (21).

11. CHOPPER METHODS

Because modern operational amplifiers drift very little with temperature and power supply variations and with time, there is a decreasing need to employ the chopped light beam principle. However, in cases where a system is to work in the presence of some ambient illumination at the detector, light·chopping is still useful. In such cases, care must be taken not to overload the detector or electronics, otherwise non-linearity might result, or at worst, saturation.

If a chopped beam is desired, this is most usually attained mechanically, but if the light source is appropriate (such as a solid-state, light-emitting diode or a small discharge lamp) then purely electrical pulsing may be employed. In either case, a synchronising signal may be extracted from the chopper and applied to the reference channel of a phase-sensitive rectifier. This technique results in an output which is proportional only to that portion of the detector current which results from the light beam, and ignores that due to the ambient illumination.

In its most developed form, this is the principle of the *lock-in amplifier*, which also has the attribute of resolving a signal which is apparently buried in noise.

12.   THE LOCK-IN AMPLIFIER[16−18]

Figure 20 shows the basic layout of a system using a phase-sensitive detector (p.s.d.). Here, for the purposes of illustration, a weak and noisy light signal is chopped via a chopper wheel, detected by a solid-state photo-sensor, and amplified. At the same time, the beam from a light source (such as a filament lamp or light-emitting diode (LED)) is chopped by the same

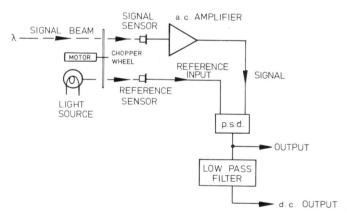

Fig. 20.   Basic phase-sensitive detector system.

wheel, and applied to a second detector. This produces a synchronising signal called the reference input, which is applied to the p.s.d. along with the main signal, as shown.

Figure 21 shows how the chopped signal is operated on by the p.s.d. Figure 21(a) represents the chopped but noisy signal produced by the photo-sensor, and this is essentially unidirectional. On passing through the a.c. amplifier, the signal itself becomes a.c., with its mean level defining the x-axis, as shown in Fig. 21(b).

Now suppose that the function of the p.s.d. is to pass each positive-going half-wave without modification, but to invert each negative-going half-wave. This is equivalent to multiplying each positive-going half-wave by +1, and each negative-going half-wave by −1, and a signal appropriate for this purpose may be derived from the reference input as shown in Fig. 21(c). The product of waveforms (b) and (c) is shown in Fig. 21(d), where an essentially reconstituted d.c. signal appears.

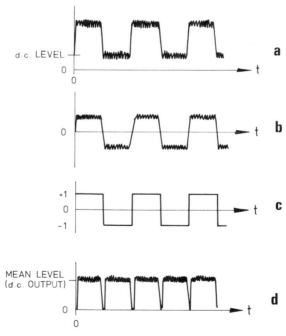

Fig. 21. Phase-sensitive detector waveforms. (a) Signal output from sensor; (b) signal output from a.c. amplifier; (c) p.s.d. gating waveform derived from reference input; (d) output from p.s.d.

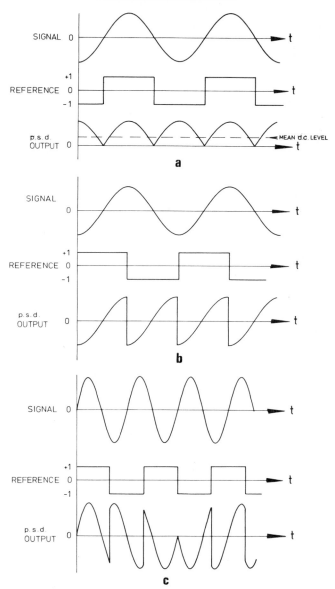

Fig. 22. Phase-sensitive rectifier outputs with (a) signal and reference in phase; (b) signal and reference 90° out-of-phase; (c) signal and reference at different frequencies. (NB, zero d.c. level in cases (b) and (c).)

At this point, it is reasonable to ask precisely what has been accomplished by such signal processing, especially when it is noted that half of the weak signal has been lost in the chopper anyway. The answer is that, provided the reference input is exactly in phase with the (noisy trapezoidal) signal input, then the waveform of Fig. 21(d) can simply be smoothed by a low-pass filter, and the noise will be largely averaged out.

This situation can easily be seen graphically by taking a simple sine wave input as in Fig. 22, and observing what happens when (i) the reference input is in phase with it; (ii) when they are out-of-phase; (iii) when they differ in frequency.

In case (i), a simple full-wave rectification takes place with a resulting d.c. level. In case (ii) the d.c. level reduces as the phase difference increases. Finally, in case (iii), there is no discernable d.c. level. Case (iii) can be extended by noting that noise is random, so that it cannot theoretically contribute to the d.c. level at all. Also, case (ii) indicates that a signal at the fundamental or reference frequency will be extracted from its own harmonics by the system.

Finally, the contribution of the dark level is automatically removed by the presence of the a.c. amplifier.

(As a practical note, the chopper should not run at mains frequency or a multiple of it, otherwise mains interference will not be properly rejected.)

13. CONCLUSION

It is hoped that this chapter has given a basic insight into some contemporary methods of radiation sensing and signal processing. The treatment has been at a fairly elementary level, but has sufficed to indicate that modern solid-state detectors can usually replace vacuum devices; and that some sophisticated signal processing methods can be utilised without too much reliance on purpose-built electronics.

Inevitably, much has been omitted. For example, other detectors do exist, such as the essentially mechanical Golay cell, or the recently-developed pyroelectric sensor for infrared applications. Also, there is a very considerable body of work relating to signal processing methods involving correlation and convolution techniques, and the recent solid-state products such as charge-coupled devices (c.c.d.'s) which make them possible. However, until commercial products (like lock-in and box-car amplifiers) become common, these more sophisticated methods must involve the cooperation of the electronic engineer.

REFERENCES

1. Silonex Inc., formerly National Semiconductors Ltd (Montreal). *Data Sheet NSL-367.*
2. Gage, S., Hodapp, M., Evans, D. and Sorensen, H. *Optoelectronics Applications Manual*, 1977, Hewlett-Packard and McGraw-Hill, New York, chapter 4.
3. Watson, J. *Semiconductor Circuit Design*, 3rd Edn, 1977, Adam Hilger and Halsted Press, New York.
4. Edgerton, Germeshausen & Grier Inc., *Data Sheet D3003B-2* and *Application Note D3000C-1.*
5. Integrated Photomatrix Ltd, *Information Sheet No. 205/A.*
6. Kelly, B. O. Lateral-effect photodiodes. *Laser Focus Magazine*, March, 1976, 38–40.
7. Petersson, G. P. and Lindholm, L. E. Position-sensitive light detectors with high linearity. *IEEE J. Solid-State Circuits*, **SC-13**(3), June, 1978, 392–9.
8. Dyck, R. H. and Weckler, G. P. Integrated arrays of silicon photodetectors for image sensing. *IEEE Trans. Electron Devices*, **ED-15**, April, 1968, 196–201.
9. *IEEE J. Solid-State Circuits*, Special issue on optoelectric devices and circuits, **SC-13**(1), February, 1978.
10. Tait, J. M. Design & application of silicon photodetectors. *New Electronics*, 25 July, 1978, 27–31.
11. Graeme, J. Op. Amp. boosts phototransistor speed. *Electronic Design*, **5**, March, 1972, 62.
12. Edgerton, Germeshausen & Grier Inc., *Data Sheets D30078-1, D3009 B-1* and *Application Note D3011B-1.*
13. Centronic, *Information Sheet on Series OS1-5.*
14. Agourdis, D. C. and Fox, R. J. Transresistance instrumentation amplifier. *Proc. IEEE*, **66**(10), October, 1978, 1286–7.
15. New, B. M. Versatile electrooptic alignment system for field applications. *Applied Optics*, **13**, April, 1974, 937–41.
16. Danby, P. C. G. Signal recovery using a phase-sensitive detector. *Electronic Engineer*, January, 1970, 36–41.
17. Brower, R. Taking noise out of weak signals. *Electronics*, 8 July, 1968, 80–90.
18. Ortec Brookdeal (Bracknell, Berkshire/Oak Ridge, Tennessee). *Guide to lock-in amplifiers.*

Chapter 2

PHOTOELASTIC TRANSDUCERS

D. R. MOORE

AVT Engineering Services, Stockport, UK

NOTATION

E Modulus of elasticity
K Dimensionless constant representing the optical strain sensitivity of the birefringent material
N Number of fringes
k Ratio of principal stresses
l Length of light path
ε_x Major principal strain
ε_y Minor principal strain
λ Wavelength of light
μ Poisson's ratio
σ_x Major principal stress
σ_y Minor principal stress

1. INTRODUCTION

The basic principles of photoelasticity have been known for many years.[1] This chapter deals with how they are used in transducers to measure strain, load and stress. The photoelastic effect is dealt with briefly in order to explain what is happening within the transducers. The linear relationship between fringe order and strain or stress in the birefringent material is emphasised.

39

The well-known use of photoelasticity in models[2] is not considered as a transducer application, but the photoelastic coating technique where components are coated with sheets of birefringent plastic can be considered as a transducer application.[3]

Photoelastic transducers for measuring load and pressure[4] are not as well known as they should be. Although their accuracy is not as high as other methods, such as those using electrical resistance strain gauges, their robustness and long term reliability render them particularly useful in civil and mining engineering applications. High modulus inclusion gauges respond to the change in stress in the host material which is useful for non-linear material behaviour. The photoelastic high modulus inclusion can be made of either plastic or glass and each acts as a stress indicating transducer in well defined modulus ratios for the host materials.[5,6]

Low modulus inclusion gauges respond to the change in strain in the host material. Plastic photoelastic low modulus transducers[7] can be used to measure strain changes in a wide range of materials and, as with the high modulus gauges, they are biaxial in their response.

Finally, a simple method of using a photoelastic transducer as a strain gauge is described.

2. PHOTOELASTICITY

When a beam of polarised light is passed through a stressed birefringent material, the beam is split into two components. These two components are in the planes of the two principal strains. Both components are retarded as they pass through the birefringent material. The relative retardation between the two components, α, is proportional to the difference in magnitude of the two principal strains and the length of the light path through the material.

$$\alpha = Kl(\varepsilon_x - \varepsilon_y) \tag{1}$$

Because of this retardation the two emergent beams will interfere with each other when combined in an analysing polarising filter. When white light is used this results in colour fringes being produced, the colour depending on the value of α. The colour will change through the spectrum as α changes by an amount λ equal to the wavelength of $0\cdot576\,\mu m$. As α changes to $N\lambda$ where $N = 1, 2, 3\ldots$, etc., then the spectrum repeats. The colour at the integral fringe order is the change between red and blue. Hence

$$\varepsilon_x - \varepsilon_y = \frac{0\cdot576\,N}{2Kl} \tag{2}$$

Where monochromatic light is used, λ is the wavelength of that light and extinction occurs when N is an integer, and black fringes will occur against a background of the colour.

The above is true for the crossed polariscope which is used for all the instruments described in this chapter. By identifying the fringe order, N, at any point, the strain difference (which is proportional to the shear strain) at that point can be determined.

At points, or areas, where a direction of principal strain coincides with the direction of polarisation, the light passes through without being affected. In the crossed polariscope these areas will appear black and are called isoclinics. This enables the direction of the principal stresses to be determined. If quarter wave plates are incorporated in both the polariser and analyser then circularly polarised light is produced. With this there is no specific direction of polarisation and therefore no isoclinics will be produced. By using removable quarter wave plates the stress directions at a point can be determined by viewing the isoclinics without the plates and then the isochromatics can be studied to determine the shear strain when the plates are replaced.

With many of the photoelastic transducers the system is simplified. First the direction of principal strain may be known or recognisable from the pattern symmetry. In this case a fixed quarter wave plate may be used. Secondly the pattern produced may be of a definite shape and it is only necessary to take readings at one particular position.

3. PHOTOELASTIC COATINGS

The most widely used application of photoelasticity as a transducer is the coating technique.

A layer of birefringent plastic is bonded to the surface of the component to be tested. The thickness of this layer must be controlled as the sensitivity is proportional to the length of the light path. In this case the length of the light path is twice the thickness, t, of the coating layer as polarised light is reflected off the back of the coating—usually by using a reflective adhesive.

$$\varepsilon_x - \varepsilon_y = \frac{N}{2tK} \qquad (3)$$

Flat sheets of plastic of controlled thickness and precalibrated sensitivity can be obtained from manufacturers for use on flat surfaces. These can be cut and filed to the required shape and then bonded.

Where curved surfaces are to be studied then the plastic must be contoured to fit the surface accurately. This is achieved by pouring a liquid resin/hardener mix into an open, flat and level mould. The liquid spreads into a layer of even thickness. After a short time the liquid semi-polymerises into a rubbery consistency which can be lifted from the mould, cut to shape with scissors and formed on to the component. After 24 h it will have fully cured and can be removed for measuring the thickness prior to bonding.

The principle is based on the plastic following the change in strain in the surface of the component. Because of this the bond between the plastic and the component must be good and care must be taken in surface preparation to ensure a good bond. The adhesive must not itself be birefringent. To avoid this and also give a surface for reflecting the light, the adhesive is usually filled with aluminium flake.

It can be seen from eqn (3) that the sensitivity of the plastic depends on the values of thickness, t, and strain optical sensitivity, K. For ease of analysis it is best to work with fringe orders between 1 and 4. Below half fringe the colours are changing from black to light grey and it is difficult to obtain the fractional fringe order accurately. Above 5 fringes the colours become fainter and the fringes may be so close together as to need magnification to analyse them.

Plastics available commercially have values of K varying from 0·02 to 0·16. The plastics with the lower K factor are low modulus polyurethanes and are used where high strains can be expected, as on rubbers or plastics. Those with the higher K factors are epoxies and polycarbonates, and are used on higher modulus materials such as metals and concrete. Where strain levels are known approximately, the plastic with a K value and thickness to give between 1 and 4 fringes can be selected. When the strain levels are completely unknown, either different thicknesses of plastic may be used on symmetrically similar areas, or similar components; or a first coating can be stripped off and replaced by a different thickness coating for a re-run of the tests.

Figure 1 shows schematically a reflection polariscope for reading the fringe pattern in the plastic coating. A white light source is normally used, but there will be provision for using a monochromatic filter. The quarter wave plates can be removed either physically or by rotating through 45° so that they are not effective. The polariser and analyser can then be rotated together so that the isoclinics can be observed and the stress directions plotted. On replacing the quarter wave plates, the isochromatics can be studied to determine the magnitudes of the principal stress difference.

The system of determining the fringe value, N, is excellently described in

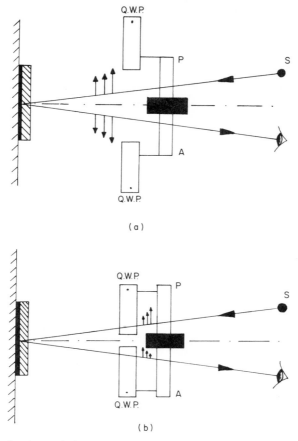

Fig. 1. A reflection polariscope. (a) Plane polarised light; (b) circular polarised light.

ref. 3. At a chosen position the value of the nearest whole number of fringes is determined and then the fractional fringe order is determined by the Tardy compensation technique; in this, rotation of the analyser causes the nearest fringe (red to blue boundary) to move towards the position. The angle of rotation required to do this, expressed as a fraction of 180°, gives the fractional fringe order. The direction, clockwise or anticlockwise, which the analyser has to be rotated to move the fringe to the position assists in the assessment of whether the principal stress in the direction of viewing is tensile or compressive.

Photoelastic coatings show the full field stress distribution over the coated area. The stress directions and principal stress difference at any point can be determined. At an edge the principal stress normal to the edge is zero and therefore the principal stress can be measured. As most failures occur at the free boundaries of holes or notches, etc., this is a very useful facet of the technique. Its advantage over alternative methods of surface stress measurement such as electrical resistance strain gauges is, therefore, that no positions of high (or low) stress are missed and strain gradients are readily seen and plotted.

The accuracy of the photoelastic coating depends on the thickness of the coating (t) and the sensitivity (K). With a coating of thickness 3 mm and K factor 0·1 it is possible to determine the magnitude of the maximum shear strain ($\varepsilon_x - \varepsilon_y$) to $\pm 10 \times 10^{-6}$ mm/mm.

Measurements may have to be corrected in analysis due to errors caused by the reinforcing effect of the plastic and due to the fact that there is a strain gradient through the thickness of the coating when the component being treated is subjected to bending stress (Fig. 2).

The plastic may also have some initial colour pattern in it prior to loading. This would be caused by mishandling of the plastic during preparation and bonding. This should not occur, but if the coating cannot be replaced corrections can be made.

Temperature changes will result in differential expansion of the coating and the component. This will strain the plastic, but as it will do so equally in all directions, $\varepsilon_x = \varepsilon_y$ and $\varepsilon_x - \varepsilon_y = 0$ so that this effect produces zero birefringence. At the edge, however, ε_2 will be zero and a fringe pattern will occur, probably to four times the thickness from the edge. Temperature effects will cause errors if a technique known as oblique incidence is used to

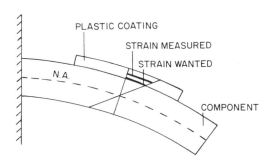

Fig. 2. Error in reading of bending strains.

determine individual principal stresses. Where possible, the test temperature should be the same as that at which the bonding of the plastic was
carried out.

The systems of correcting for the above errors are well documented in
ref. 8.

Although most photoelastic coating studies are concerned with
measuring principal stress difference, a system known as oblique incidence
is available for determining individual principal stresses.[3] With this, the
incident light is not normal to the surface of the plastic, but oblique to it in
the direction of one of the principal stresses. By choosing a suitable angle
for the incident light, the principal strains ε_x and ε_y can be easily calculated.
The apparatus used is somewhat clumsy if hand held and only tripod
mounted instruments will give good accuracy. However, it is often useful to
know the principal strains, even to an accuracy of 15 %. A particular case
would be where $\varepsilon_x \simeq \varepsilon_y$ and it is required to know if ε_x and ε_y are large or
small.

The applications of photoelastic coatings are numerous. It is probably
the only experimental stress analysis technique which gives an overall
picture of the stress distribution and can also be used for accurate point to
point measurements.

Not only are areas of high stress readily shown, but also areas of low
stress, where thickness reductions can be made—saving cost and weight.
Areas of uniform stress are shown and also areas of high stress gradients.

Although electrical resistance strain gauges can be individually more
accurate than photoelastic coatings, they can also be in the wrong place and
the total knowledge given by the coating would therefore be the more
accurate.

4.   PHOTOELASTIC LOAD CELLS

When a disc of birefringent material is loaded across a diameter and viewed
with polarised light, fringe patterns occur as shown in Fig. 3. There is a
concentration of fringes at the contact points, but across the other diameter
there is a simpler fringe pattern. This increases in fringe order in a linear
relationship with applied load.

The stress concentration at the contact lines is such that plastic cylinders
will creep under load. However, glass behaves elastically almost up to
fracture and is always used in photoelastic load cells.

The load required to produce one fringe at the centre of a cylinder of

D. R. Moore

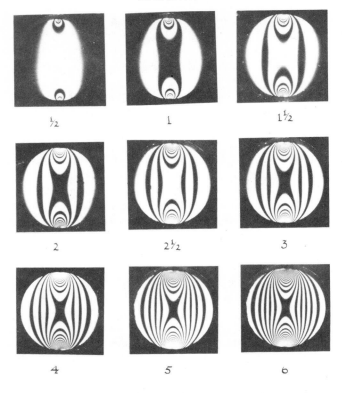

FRINGE COUNTING

Fig. 3. Load cell fringe patterns.

glass depends on the diameter of the cylinder, the wavelength of the light used and whether the light is transmitted straight through the glass or reflected off the back surface so that it passes through the glass twice. The sensitivity of glass cylinders of different diameters when using red light is given by Table 1.

This table is based on using red light. Monochromatic light is used because it is easier to discern the black fringe on a coloured (red) background than to observe the red to blue boundary of fringe orders greater than four. In practice values up to seven fringes are used in load cells.

If the glass disc were to be used on its own as a load bearing transducer, it

TABLE 1

Glass diameter (mm)	Load per fringe (newtons)	
	Transmitted light	Reflected light
50·0	4 715	2 355
25·0	2 355	1 180
12·5	1 180	590

would obviously limit the range of the system and also create practical stability problems.

In photoelastic load cells the glass transducer element is placed in a hole in a steel body between steel platens held in diametrically opposite slots. As load is applied to the steel body the hole will tend to deform and this deformation is registered by the glass cylinder.[9]

Figure 4 shows a diagrammatic arrangement of a photoelastic load cell using transmitted light. Also shown is the mounting platen arrangement. The top mounting platen is a simple section of flat steel. The bottom mounting platen is a sliding wedge system. This wedge system facilitates assembly of the glass cylinder and can be adjusted during calibration.

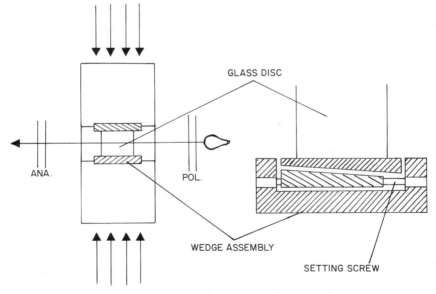

Fig. 4. Transmitted light load cell and wedge mounting platen.

With direct compression load cells the steel body is generally cylindrical and loaded across the ends. The glass element is fitted in a hole drilled diametrically across the centre of the cylinder and transmitted light is used. The sensitivity of the load cell can be adjusted in design by arranging the dimensions of the steel body and the glass transducer element.

As with all load cells it is essential to ensure axial loading and this can be done by using spherical or roller bearings on the ends.

It is often necessary to measure the loads in tension rods, rock bolts or wire tendons. In this case the end of the rod or tendon passes through a hole drilled along the axis of the load cell body. The load cell is then held in compression against the bearing plate. The glass element cannot therefore be at the centre of the diametric hole, but is offset to the side of the axial hole. The back of the glass is reflected, probably by an aluminium spray, and reflected light viewing is used.

The direction of loading across the glass cylinder is known and, therefore, so is the stress direction at the central reading position. There is no requirement for viewing isoclinics to determine stress directions and a simple, inexpensive, hand held polariscope can be used for taking readings. Figure 5 shows a viewer for reflected light transducers. A torch is used for a light source at one side of the viewer. Holding the handle in the direction of

Fig. 5.   Precision load cell viewer.

the loading platens ensures correct alignment of the instrument. With the scale set at zero the number of whole fringes is counted and rotation of the scale enables the fractional fringe order to be determined.

By this means the fringe value to 0·02 fringes can be determined. With a standard range of six fringes resolution is therefore to 1 part in 300 of the full load range. Hence small changes in load can be readily discerned.

Despite the use of ball or roller bearings, photoelastic load cells may be subject to small errors due to service misalignment of the loading surfaces which can be encountered in civil and ground engineering. Due to the difference in values of coefficient of expansion of glass and steel, thermal errors occur. The value of the thermal error varies according to the dimensions of the load cell, but can be easily calculated.[9] However, in most applications the error is small enough to be ignored.

Precautions have to be taken to avoid corrosion of the platens which hold the glass as this would cause long term errors. 'O' ring seals are used to prevent ingress of moisture and special steels are used to prevent electrolytic action with the glass.

Where reasonably controlled loading conditions apply, the overall accuracy of photoelastic load cells is better than $\pm 1\%$ of the full load range.

For particular applications where a high degree of accuracy has been required over a narrow band within full load, a special technique has been developed. This was first used for a concrete prestressing load cell where it was required to know when a given load had been reached in the stressing tendon. This load would vary, within limits, according to the design of concrete beam. When applying the prestress it is not necessary to know the existence of low or high loads, but only the particular design load.

A cell was designed to have a safe working load of 800 kN. The design of the cell, however, was such that a load of only 60 kN would produce six fringes. A screw mechanism moves the wedge platen arrangement so that it is not in contact with the glass disc, which is held in position by silicone rubber.

The load cell is placed in a calibration machine and brought to the required load. The wedge screw mechanism is now used to bring the platen in contact with the glass and it is tightened further so that the second fringe order appears. An easily recognisable fringe pattern now occurs when the load reaches the required prestressing value. The measuring load range is only ± 30 kN around the required value of 800 kN, ensuring greater accuracy.

Photoelastic load cells use glass and steel loaded well within the elastic limit. So long as the glass has been aged for a year before use, creep in the glass is negligible and is partially compensated by creep in the steel. One load cell has been held for 14 years under constant load with no measurable change in the fringe pattern. The cells are, therefore, recommended for use where long term stability under sustained load is required.

The readout system is entirely separate from a photoelastic load cell. There are no vulnerable hydraulic or electrical connections. Hence they are ideal for use in the arduous conditions encountered in civil and mining engineering applications.

With one exception, mentioned later, the cost of a photoelastic load cell is comparable with that of an electrical resistance gauge or a vibrating wire gauge load cell. The cost of the readout instrument, the 'Precision Hand Viewer', is far less than that of the readout systems for electrical resistance

Fig. 6.   Rock bolt tension indicator.

and vibrating wire gauges. Where only a small number of load cells is needed, the saving on readout instruments can be a decided economic advantage.

The glass sensing element of the load cell can only be damaged by direct and severe impact as the glass is generally over 12 mm thick. Such damage is immediately evident and erroneous readings cannot be taken.

The disadvantages of photoelastic load cells are that remote readings can only be taken by the use of expensive television camera systems and they are not suitable for automatic data logging.

The exception on price comparability mentioned above is the Rock Bolt Tension Indicator (RBTI). This is an inexpensive transducer for monitoring the load in rock anchor bolts of up to 22 mm diameter. The RBTI, shown in Fig. 6, has a sensitivity of 10 kN per fringe and a range of 60 kN. The RBTI can be calibrated, using the wedge platen system, to read any 60 kN range between 0 and 180 kN. Spherical washers are used to ensure axial loading and the accuracy is claimed to be better than ±0·5 kN.

Similar transducers are also used to measure pressure. In this case a piston is used to load on to a glass disc. The range of the hydraulic pressure photoelastic transducer is determined by the diameter of the piston and the diameter of the glass disc.

5. PHOTOELASTIC HIGH MODULUS INCLUSION GAUGES

The theory of the high modulus inclusion gauges has been considered by many authors.[5,10,11] In essence, if the inclusion gauge is sufficiently rigid in comparison with that of the host material, then the strains induced in the gauge by stress changes in the host material are relatively small. The stress changes in the gauge are therefore independent of the modulus (stress/strain behaviour) of the host material. If the gauge is not a rigid inclusion, its response to stress changes in the host material will need to be calibrated for all variations in the value of modulus.

In many materials the modulus changes radically according to such factors as temperature, rate of loading and intensity of stress. In order to measure change in stress in such materials it is essential to use a high modulus inclusion gauge.

Hawkes and Fellers[5] have shown that a glass photoelastic stress gauge acts as a high modulus inclusion in many rocks, concrete and ice. The plastic photoelastic stress gauge[6] acts as a high modulus inclusion gauge in low modulus materials such as rubbers and soft polymers.

5.1. The Glass Photoelastic Stress Gauge

This gauge consists of a hollow cylinder of glass with an outer to inner diameter ratio of 5:1. Typical dimensions of the gauge are 30 mm outer diameter, 6 mm inner diameter and 40 mm long, although these may be varied considerably for special applications.

The gauge is bonded into a hole drilled into the rock or concrete or it can be cast directly into concrete as it is poured. The cement used to bond the gauge into the hole must be cold setting, cure in an acceptable time, set and bond in moist conditions and have a negligible effect on the response of the gauge. A heavily filled epoxy resin cement is used and the diameter of the hole should be such that the annulus of cement is not greater than 3 mm.

It is necessary to place the gauge in the hole, in the required location, with an even annulus of cement. The face of the gauge towards the entrance of the hole must be clear for subsequent viewing. A variety of setting tools are available for use in different lengths of holes. Gauges have been set at the back of holes up to 14 m long.

Figure 7 shows a glass stress gauge. The hollow aluminium cylinder at the back of the gauge terminates in a reflector. The light probe contains polarising filters so that polarised light is transmitted through the gauge. A similar instrument to that shown in Fig. 5 is used for viewing the fringes.

The pattern seen in the gauge depends on the magnitude of the major principal stress and the ratio between the major and minor stresses. Typical fringe patterns are shown in Fig. 8. Analysis of the fringe patterns requires some experience, but is not difficult provided the basic rules are followed.

The first factor is to decide whereabouts to take the reading. In selecting these positions two important factors have been considered: (i) variations

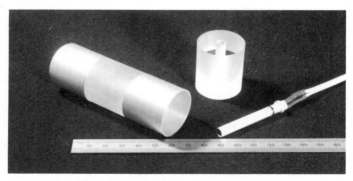

Fig. 7. Glass photoelastic stress gauges and light probe.

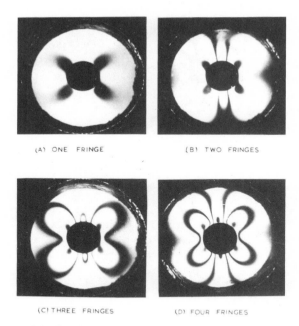

(A) ONE FRINGE   (B) TWO FRINGES

(C) THREE FRINGES   (D) FOUR FRINGES

Fig. 8. Stress gauge (biaxial) fringe patterns.

in the observed fringe order with applied stress should be as large as practicable to attain maximum sensitivity; (ii) the rate of variation in the fringe order along the radius of the gauge should be as small as possible so that any errors in measurement due to lack of precision in locating the observation points do not cause large changes in the calculated stress values.

From theoretical analysis and experimental evidence the points along the $0°$, $45°$ and $90°$ axes at a distance from the centre of $1·36$ times the radius of the central hole have been chosen. These are shown in Fig. 9. The light probe contains a collar of the requisite diameter so that the readings can be taken at the collar edge.

The axes of symmetry of the fringe pattern are always in line with the principal stress directions in the rock body. The patterns in Fig. 8 all have the numerically greater stress in the vertical direction.

The sign of the major principal stress is determined by aligning the viewer with the direction of the maximum stress and rotating the scale clockwise. If the fringes along the $90°$ axis move towards the centre, then the maximum

D. R. Moore

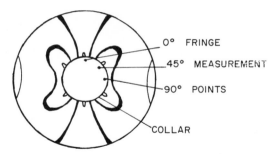

Fig. 9.   Optimum measuring positions.

stress is compressive. If they move outwards then the maximum stress is tensile. This is because the polarising filters are aligned in the opposite way to that of the polariscope used for the reflective coating technique.

With experience the ratio k, between the major and minor principal stresses, can be obtained by pattern recognition. At best, however, it is only possible to distinguish between $k = 0, \frac{1}{4}, \frac{1}{2}, \frac{3}{4}$ and 1, provided the fringe value is greater than 1·5. To obtain more accurate and reliable information the following method is used:

(i)   With the viewer scale set at the zero position, determine the number of whole fringes at the collar edge at the $0°$, $45°$ and $90°$ axes. Then use Tardy compensation to obtain the fractional fringe order values.

When determining the fractional fringe order at the $0°$ and $90°$ axes, the handle of the viewer should be aligned with the maximum principal stress axis. For the fractional fringe order at the $45°$ axis the viewer axis should be turned until clear fringes are observed at the collar edge throughout the rotation of the viewer's analysing filter. As a guide, the viewer axis should be in line with the maximum stress axis when the pattern suggests $k = 0$ and change progressively to $45°$ when $k = 1$.

(ii) The ratios between the fringe orders at (a) $0°$ and $90°$ and (b) $45°$ and $90°$, are then calculated. Using the $0°/90°$ ratio in conjunction with Fig. 10(a) two possible values of k are obtained. The $45°/90°$ ratio is then used in conjunction with Fig. 10(b) to determine which value of k is correct.

The sensitivity of the biaxial photoelastic stress gauge is defined, in terms of the maximum principal stress, as the stress per fringe per unit gauge length. Its value is governed both by the ratio of the modulus of the gauge to that of the rock E_i/E_h, and the ratio of the minor to major principal stress, q/p.

If the E_i/E_h ratio is > 2.5 the gauge sensitivity in a particular stress field remains sensibly unchanged. If the modulus of the rock, or concrete, is higher, then for accurate measurement the gauge should be calibrated in the particular rock. Otherwise a gauge sensitivity of 105 N mm^{-2} per fringe per mm gauge length can be used in a uniaxial stress field for readings taken on the 45° axis. For a gauge 25·4 mm long the sensitivity is 4·14 N/mm^{-2} per fringe.

Using the gauge sensitivity (S_0) for the uniaxial stress field, the gauge sensitivity (S) for the major principal stress in a biaxial stress field can be obtained by multiplying S_0 by an appropriate coefficient obtained from Fig. 11. The major principal stress is then calculated as $p = SN/l$ where N is the fringe order and l is the length of the gauge. The minor principal stress, q, is then calculated using the previously determined stress ratio, k.

Hence a relatively inexpensive gauge can be used to obtain the direction, sign and magnitude of the biaxial stress field in the diametrical plane of the gauge. Readings can be reliably taken over a long period. Gauges set 8 years ago in a salt mine are still being used.

In most circumstances they respond to the change in stress from the time they are installed. However, in materials which exhibit creep characteristics, e.g. rock salt and frozen ground, the *in situ* stress is, in time, transferred to the high modulus gauge. By plotting a graph of stress against time a value of the *in situ* stress is obtained. The time taken to accept the *in situ* stress depends on the viscoelastic nature of the host material. It has been shown that in the mass concrete of a new dam, the time is approximately 1 year.[12]

As with the photoelastic load cell, the disadvantages of the stress gauge is that remote readout and automatic recording are not feasible. Visual access, in some circumstances, may only be available during construction. In such circumstances stress gauges can be used in conjunction with vibrating wire gauges. By using the two together during the construction stage, a realistic value of modulus can be obtained—and the vibrating wire can truly be said to have been calibrated in site conditions.

5.2. The Plastic Photoelastic Stress Gauge

Glass is too insensitive to measure the low stresses which need monitoring in soft polymers. Gauges made from epoxy resin or polycarbonate have been developed for work in these materials.

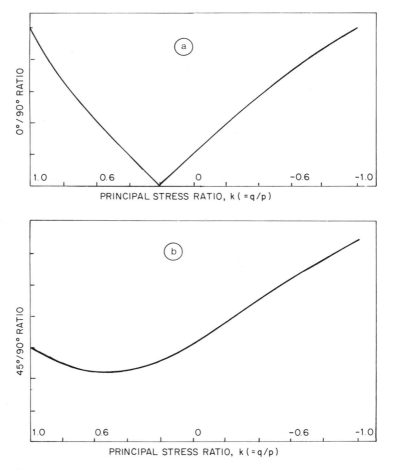

Fig. 10.   Relationships between the fringe order ratios and the principal stress ratio.

As they are normally used in comparatively small components, the gauges are made smaller than the glass ones. A typical size is 15 mm long by 5 mm diameter and with a 1 mm diameter central hole.

Because of their size the viewing system requires a magnification system. A reflection polariscope shown in Fig. 12 is used. With this a beam of polarised light is refracted through a prism and then reflected off a small front surfaced mirror so that it enters the gauge. It reflects off the back of

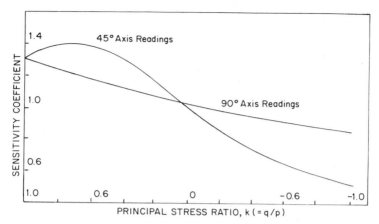

Fig. 11. Relationship between gauge sensitivity and principal stress ratio.

the gauge and passes round the mirror into a $40\times$ magnification telemicroscope. The telemicroscope is split by the analyser so that the fringe pattern is viewed through the eyepiece. Because of this construction isoclinics can be used for aligning the polariscope for the $45°$ position readings.

As the light is being passed into and out of the gauge on the same line there is a danger of front face reflection obscuring the fringe pattern. This is avoided by putting a $2°$ angle on the front face of the gauge to deflect this reflection away from the objective lens.

Placing a reading collar in such a small gauge is difficult. The centre of the gauge is therefore countersunk to the correct diameter and left unpolished to act as the collar.

Great care must be taken during manufacture of the gauges to avoid inducing parasitic birefringence. Great care is needed during machining and the faces of the gauges have to be hand polished.

The gauges are usually cast into the polymer during manufacture of the components. They are held in position by pins passing through the centre hole. Nylon gauge holders, the same diameter as the gauge, keep the gauge face clean and when removed leave a suitable viewing hole.

The reading system is exactly the same as for the glass stress gauge, although because the Poisson's ratio of plastic is different from that of glass, the reference graphs do not follow the same line. The sensitivity of a $15\cdot8$ mm long epoxy resin gauge is $132\cdot5\,\text{kN m}^{-2}$ per fringe.

Both polycarbonate and epoxy resins absorb moisture. Whilst absorbing

Fig. 12. Polariscope for plastic stress gauges.

moisture spurious fringe patterns occur, although when equilibrium of 'saturated' resin is reached the spurious fringe disappears. In tests where the humidity conditions vary it is essential to keep the gauge dry by using molecular sieve in the viewing access hole between taking readings.

The photoelastic plastic stress gauge responds to stress changes and not strain changes. In low modulus non-linear viscoelastic materials no account need be taken of the modulus changes which occur due to loading rate, temperature or stress level changes.

6. PHOTOELASTIC STRAIN GAUGES

Figure 13 shows a photoelastic strain gauge. It consists of a rectangle of photoelastic plastic into which a fringe pattern has been frozen. The pattern consists of coloured lines which go across the short axis of the gauge. A

Fig. 13.   Photoelastic strain gauge.

combined linear and quarter wave polarising filter covers the top of the gauge so that the fringes can be seen merely by shining a light on to the gauge. The back of the gauge is coated with a reflective paint.

Behind the reflective coating is a piece of vinyl tape which covers all but the last 6 mm of each end of the gauge. This serves two purposes. It protects the reflective layer from scratches and does not bond to the adhesive used to fix the gauge.

In use the gauge is bonded by its ends to the component. For most materials a room temperature setting epoxy or polyester cement may be used for this purpose. When strain is applied to the component the coloured fringes move along the gauge. This movement can be measured on the scale at the side of the gauge. Tensile strains move the fringes to a higher scale number and compressive strains move them to a lower number.

The movement of a fringe is directly proportional to the axial strain along the gauge. The gauge shown in Fig. 13 has a sensitivity of 50×10^{-6} strain per division. This is not as accurate as alternative strain gauges such as electrical resistance or vibrating wire types. There is a use for the gauges, however, in the hazardous environments sometimes found in mines or chemical plants where electronic instruments may be forbidden. Also, as they can be quickly and cheaply installed—some adhesives taking 5 min to cure can be used—they can be useful for rapid assessment of strain change levels. In many cases the gauge can be removed and reused.

REFERENCES

1. Frocht, M. M. *Photoelasticity*, vol. 1 (1941) and vol. 2 (1948), John Wiley, Chichester.
2. Zienkiewicz, O. C. and Holister, G. S. Three-dimensional photoelasticity, chapter 13 in: *Stress Analysis*, 1965, John Wiley, Chichester.

 3. Photolastic Inc. (Malvern, Pennsylvania). Introduction to stress analysis by the photoelastic coating technique. *Tech. Bulletin TDG-1*, 1974.
 4. Hawkes, I., Dhir, R. K. and Rose, H. *Civ. Eng.*, **59**, 1964, 1536–40.
 5. Hawkes, I. and Fellers, G. E. *Int. J. Rock Mech. & Min. Sci.*, **6**, 1969, 143–58.
 6. Buswell, H. J., Moore, D. R. and Owens, A. *Expt. Mech.*, **14**, 1974, 274–80.
 7. Hawkes, I. *Strain*, **3**, 1967, 1–7.
 8. Photolastic Inc. (Malvern, Pennsylvania). Corrections to photoelastic coating fringe order measurements. *Tech. Bulletin T-405*, 1977.
 9. Hooper, J. A. *The theory and development of load cells incorporating photoelastic glass disc transducers.* PhD Thesis, Sheffield University, 1968.
10. Coutinho, A. *Int. Assoc. for Bridge & Struct. Eng. Pub.*, **7**, 1949.
11. Barron, K. *Mines Branch, Dept. Mines Tech. Survey*, Ottawa, 1964.
12. CIRIA (London). Instrumentation investigations in Clywedog dam. *Technical Note No. 9*, 1970.

Chapter 3

APPLICATIONS OF THE MOIRÉ EFFECT

A. R. Luxmoore

Department of Civil Engineering, University College, Swansea, UK

and

A. T. Shepherd

Ferranti Ltd, Dalkeith, UK

1. INTRODUCTION

The moiré effect is the name given to the fringe patterns observed when two regular patterns, usually linear gratings, are placed in contact, and almost in alignment (Fig. 1). The effect is commonly observed in everyday objects such as loosely woven cloth (the name originates from 'moiré silk'), pairs of railings, television, etc.

The phenomenon was first investigated by Lord Rayleigh in 1874. Righi in 1887 was the first to point out its possible use for measurement. Giambiasi in 1922 patented a caliper gauge using visual observation of the fringes, and Roberts in 1950 used a single channel photoelectric pickup and electronic counter giving total distance travelled without taking account of direction, but it is only in the last three decades that it has been used for practical measurement.

There are numerous ways in which the effect can be utilised, but its main applications to date have been in machine tool control and experimental stress analysis. If the effect was more widely appreciated there would undoubtedly be many more applications.

Consider the superimposed grids of Fig. 1. In the top pattern the gratings are parallel, but have a small difference in pitch (vernier fringes); in the lower pattern, the pitches are identical but the gratings are inclined by a

A. R. Luxmoore and A. T. Shepherd

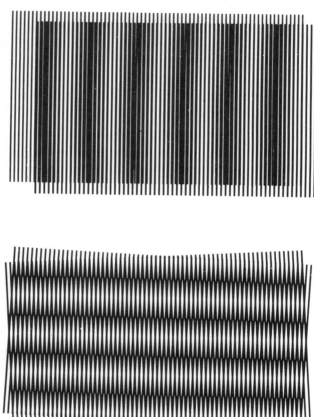

Fig. 1. Principle of moiré effect.

small angle to one another (angular fringes). In each case, the fringes are formed by the lines of one grating obliterating the spaces of the other, and this is the basic explanation of the effect in terms of rectilinear propagation of light, i.e. assuming no diffraction. In general, the fringes will be a mixture of the two types of deformation, and if the deformations vary across the grating area, then the fringes will be curved and nonparallel.

If the gratings of Fig. 1 are viewed at a distance much greater than the normal reading distance, the grating lines become less distinct, and the fringes become more pronounced. This illustrates the importance of diffraction effects in the formation of fringes, and although the grating lines may not be resolved, the fringes are still easily visible. This enables us to view fringes from gratings which are unresolved by the optical devices used to observe the moiré fringes, and this is the usual case in most applications. Observation of Fig. 1 from a large distance does not alter the geometry of the fringe pattern, and so the shape of the fringe pattern can be analysed on a geometrical basis. Diffraction effects must be considered when analysing the intensity and visibility of the fringes.

The fringes can be used to determine either the *displacement* of one grating relative to another, or the *distortion* of one to the other. The former effect is used in machine control, where movement of one grating relative to the other causes the fringes to move in a manner that magnifies the grating displacement. For example, in the angular fringe pattern of Fig. 1, if one grating moves perpendicular to the grating lines by one pitch (a line and a space), the fringes will move vertically by one fringe spacing. Similarly, in the vernier fringe pattern, a grating displacement of one pitch perpendicular to the grating lines will also cause the fringes to move one fringe spacing, but this time in the same direction as the grating displacement.

The measurement of distortion, used mainly in strain analysis, is illustrated by the vernier fringe pattern of Fig. 1. If two identical gratings are aligned parallel, no fringes are seen. If one grating is stretched perpendicular to the grating lines by, for instance, being attached to a body that is strained, then vernier fringes are formed and the spacing of the fringes decreases as the strain increases. Similarly, small changes in angle can be measured by rotating one grating relative to the other.

For both displacement and distortion measurements, the moiré effect is sensitive only to changes perpendicular to the grating lines, as displacement parallel to the grating lines will produce no detectable change in either the grating or its position.

In most applications, there is usually a fixed grating, variously described as either the reference, master or index grating, which is used to form moiré fringes with a distorted or displaced specimen grating. Moiré fringes can be formed from the two gratings in the following ways:

1. two gratings in near contact;
2. projection of one grating onto another by an optical system;
3. projection of one grating onto another via an optical medium, e.g. a mirror, liquid, etc., which can deviate the light rays and hence either distort or displace the grating image.

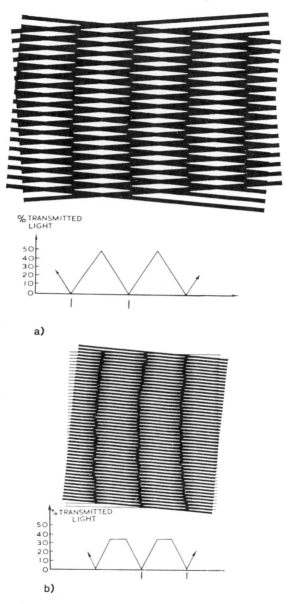

Fig. 2. Fringe sharpening effects using unequal lines and spaces and reciprocal gratings. (a) Equal lines and spaces in both gratings; (b) line/space ratio = 2:1 and 1:2. (Courtesy of G. Holister.)

2. GEOMETRICAL GRATING EFFECTS

2.1. Line/Space Ratios

The gratings in Fig. 1 both have lines and spaces of equal width. For rectilinear light propagation, which provides an adequate explanation for gratings up to 10 lines mm^{-1} (in this context, a line represents a complete pitch), light transmission takes the form of Fig. 2(a), when the transmission is averaged over several grating pitches. If the line/space ratio is changed from 1:1 to say, 1:2, and the second grating has a ratio of 2:1, i.e. one grating is the negative of the other, then the fringes are sharpened, albeit at the expense of an overall loss in fringe visibility (Fig. 2(b)). For fine gratings, diffraction effects dominate, and fringe sharpening can only be achieved by modifying the diffraction behaviour of the gratings.

The same principle can be used to produce fringes of different contrast in the same pattern.[1] In Fig. 3, every fourth bar in the top grating is twice the

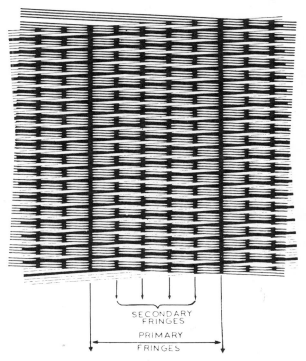

SECONDARY
FRINGES

PRIMARY
FRINGES

Fig. 3. Superimposition of positive and negative mixed grid systems, illustrating formation of dashed secondary fringes. (Courtesy of D. Post.)

width of the other bars, and when this is superimposed on its negative with a slight misalignment between the gratings, every fifth fringe is 'continuous' and hence contrasts more than the intervening 'discontinuous' fringes. For higher line densities, the individual lines are not visible, and the two types of fringes are identified by a difference in contrast. This procedure could be a useful device for fringe counting in the visual inspection of moiré fringes, but as far as the writers are aware, such gratings are not available commercially.

2.2. Nonlinear Gratings[2]

The moiré effect has been explained in terms of linear gratings, but the effect can be observed with nonlinear gratings, and this increases the range of applications. Figure 4 shows the patterns obtained when pairs of circular, radial and Fresnel zone gratings are superimposed with a small displacement between their centres. In general, the fringe pattern is complex, except for the zone gratings, where the fringes are straight and equispaced, and perpendicular to the direction of the displacement. Radial gratings are used in machine control to measure angular rotations up to 360° with a high precision, and this application will be described later. Circular gratings can be used as strain rosettes for surface strain measurements.[3]

Many other grating types are possible, and Vargady[4] discusses the uses of special grating patterns as position indicators, with one position on the specimen grating always being defined by a fringe (or fringes) of unique

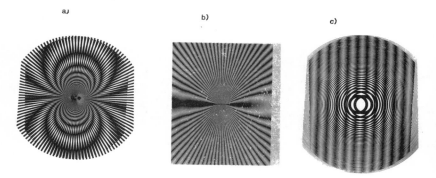

a)	b)	c)

Fig. 4. Examples of moiré fringe patterns from three pairs of nonlinear gratings. Fringes produced by superimposing gratings with a small horizontal displacement between centres. (a) Radial; (b) circular; (c) Fresnel zone.

Position of identical pitches

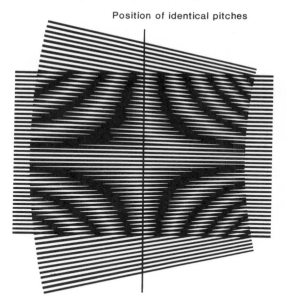

Fig. 5. Measurement of weave density with grating of linearly varying pitch.

appearance. This effect has been utilised in the textile industry for many years to measure the density of yarn in a woven fabric (Fig. 5). The cloth is represented by the linear grating onto which is superimposed a nonlinear grating which has the pitch increasing linearly with distance along the grating. The fringe pattern is in the form of a cross, with the centre of the cross located at the point where the pitches of both gratings match. Calibration of the nonlinear grating allows direct measurement of the weave density at any small area within a piece of cloth.

3. INTENSITY MODULATION OF MOIRÉ FRINGES

3.1. Modulation Produced by Coarse Gratings[5]

In Fig. 2, rectilinear propagation of light was assumed so that scattering of the light by the gratings could be ignored. However, it was assumed that the observer would still average the light transmitted over several grating lines to produce fringes with a smooth intensity contour. This effect is assisted by using fine gratings with an optical system of limited resolution, i.e. the eye. Alternatively, one may average a given area of the gratings by using the system in Fig. 6, which is the procedure used in displacement measurements.

The light transmitted by a coarse amplitude (slit and bar) grating will be directly related to the grating structure, as in Fig. 2, and can be represented by a transmission coefficient, $t_1(x)$, defined as

$$t_1(x) = \frac{\text{light intensity out, } I_0}{\text{light intensity in, } I_i}$$

where x is measured perpendicularly to the grating lines. For rectilinear light propagation, and with uniform illumination, this function will be a rectangular wave, which can be represented by an infinite Fourier series. For a grating with equal lines and spacings, this series becomes

$$t_1(x) = A_0 + \frac{4A_0}{\pi}\left(\sin\frac{2\pi x}{p} + \frac{1}{3}\sin\frac{6\pi x}{p} + \frac{1}{5}\sin\frac{10\pi x}{p} + \cdots\right)$$

where p is the pitch of the grating. When this grating is superimposed on another grating of similar pitch and transmission coefficient, $t_2(x)$, and

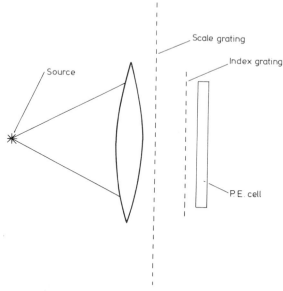

Fig. 6. Coarse amplitude gratings for measurement of displacement.

uniformly illuminated with light of unit intensity, the intensity transmitted between limits x_1 and x_2 (representing the averaged area) is given by

$$I_{02}(x) = \int_{x_1}^{x_2} I_{i2}(x) t_2(x) \, \mathrm{d}x$$

$$= \int_{x_1}^{x_2} t_1(x) t_2(x) \, \mathrm{d}x$$

Making either of the two gratings vanish outside the limits, x_1 and x_2, the integration limits can be extended to $\pm \infty$, and the intensity distribution of the fringes is given by the product of $t_1(x)$ and $t_2(x)$.

We can now distinguish between two important cases: superposition of identical gratings, and superposition of gratings with a small difference in pitch δp. For identical gratings aligned precisely, the intensity transmitted is uniform over the grating area, and any displacement between the two gratings will produce an intensity signal containing a fundamental spatial frequency of $1/p$ plus successive harmonics of considerably reduced amplitude,[7] expressed as

$$I(x) = I_0 + I_1 \cos \frac{2\pi}{p} (x - x') + I_2 \cos \frac{4\pi}{p} (x - x'')$$

$$+ I_3 \cos \frac{6\pi}{p} (x - x''') + \cdots$$

where x', x'', x''', etc., indicate the positions of the various maximum intensities for each harmonic, which should normally be the same, so that $(x - x')$ corresponds to the displacement between the two gratings.

When a small difference in pitch is present in two aligned gratings, vernier fringes are seen, and the intensity variation of these fringes is given by[8]

$$I(x) = I_0 + I_1 \cos \frac{2\pi x}{p} \cdot \frac{\delta p}{p} + I_2 \cos \frac{6\pi x}{9p} \cdot \frac{\delta p}{p} + I_3 \cos \frac{10\pi x}{25p} \cdot \frac{\delta p}{p} + \cdots$$

If the difference in pitch varies across the area of the grating, we can represent this variation as $\rho(x) = x \delta p / p$.

In most practical applications of the moiré effect, the intensity distribution is simplified so that only the fundamental term is of any importance, equivalent optically to a two beam interference system. This approach requires the gratings to be used so that their fundamental line structure is dominant.

This could be done by using gratings with a sinusoidal amplitude structure but the same effect can be achieved more simply by modifying the optical system forming the fringes. For example, projecting a grating with a lens of limited aperture will reduce the harmonics considerably, and selection of the correct aperture will allow only the fundamental to be transmitted.

3.2. Effect of Gap on Fringe Contrast

In displacement measuring applications the fringes are formed from gratings separated by a finite gap and illuminated with imperfectly collimated light which reduces the amplitude of harmonic components considerably while producing only a·slight reduction in contrast of the fundamental component.

If the collimated beam is produced from a source of width, s, by a lens of focal length, f, it will contain a smear factor proportional to the ratio s/f. If this beam is used to project the line structure of a grating of pitch p on to a second similar grating separated by a gap t a condition arises when $p/t = s/f$ whereby light leaving any point on the first grating will be smeared over a whole pitch of the second grating so that no modulation of the beam will result.

The reduced projected fringe contrast or percentage modulation of the beam with gap t can be expressed as

$$M_t = M_0 \frac{\sin \pi p f / s t}{\pi p f / s t}$$

This expression has the value unity at gap $t = 0$ and zero at gap $t = pf/s$. At a gap $t = pf/4s$ modulation is reduced to 90% of maximum and this represents a satisfactory practical limit of working gap. The effect on harmonics may be calculated by substituting p/n for p where n is the order of the harmonic.

It is necessary to maintain high fringe contrast in displacement measuring systems. The electrical signals derived from the fringes must be d.c. coupled throughout and they contain a d.c. or zero frequency component which is proportional to the mean light intensity of the fringe. This is subject to optical noise fluctuations due to random changes in grating transmission or reflection over its length or to variations in light source intensity and detector sensitivity. Although these can be minimised by operating detectors in balanced counter-phase pairs, some residual fluctuation remains. High fringe contrast ensures that the effect on measurement accuracy is reduced to an acceptable level.

3.3. Diffraction Effects

Gratings up to 10 lines mm^{-1} may be regarded as operating in accordance with geometric optics. Between 10 and 50 lines mm^{-1} diffraction effects become more significant and over 50 lines mm^{-1} diffraction effects predominate.

3.3.1. Contrast reduction with diffraction

A pair of fine slit and bar gratings in close contact will produce high percentage modulation of a light beam. When a finite gap is introduced the beam passing through the first grating is split up into a number of angular components corresponding to the orders of diffraction. Most of the energy is contained in the zero and first order spectra. The first order spectra have angular separations from the zero order of $\pm \lambda/p$ where λ is the predominant light wavelength and p the grating pitch. With silicon photocells and either tungsten or GaAs source the predominant wavelength is around 0·9 μm. The value of λ/p varies from less than 1/100 for 10 lines mm^{-1} gratings to almost 1/4 for 250 lines mm^{-1}. This diffraction smear factor causes a reduction in fringe contrast in addition to that due to the geometrical ratio s/f.

The underlying theory is much more complex and has been dealt with at length by Guild[6,7] and by Burch.[5]

For practical purposes it can be taken that fringe contrast reaches zero at a gap $t = p^2/2\lambda$ and that an acceptable reduction in modulation due to diffraction, of the order of 10%, can be maintained by operating at a gap no greater than a quarter of that giving zero contrast, i.e. $t < p^2/8\lambda$.

Contrast reduction due to diffraction is cumulative with that due to imperfect collimation and if both gap conditions are on the recommended limit the overall reduction will be 20%.

The value of maximum gap for a typical reading head with $s = 1$ mm and $f = 20$ mm is plotted for both collimation and diffraction effects against grating pitch in Fig. 7.

If it is accepted that a minimum working gap from considerations of mechanical clearance is of the order of 0·1 mm, then the finest grating pitch which is usable is between 25 and 50 lines mm^{-1}. However diffraction effects may be turned to advantage to enable the use of finer line structures.

3.3.2. Use of the Fresnel image

As stated above, a pair of slit and bar gratings of pitch p separated by gap $t = p^2/2\lambda$ produces zero modulation of a collimated length beam of

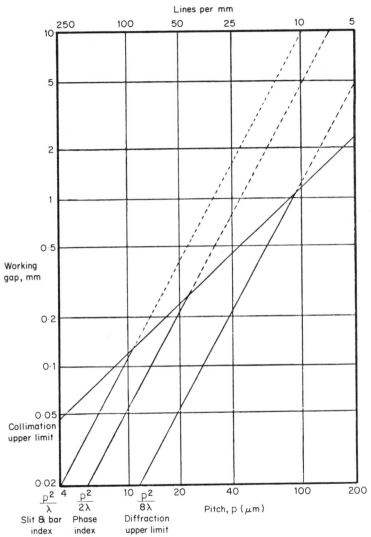

Fig. 7. Working gap versus grating pitch.

wavelength λ. However, at a gap $t = p^2/\lambda$, a high contrast image of the first grating is reconstructed and this interacts with the second grating to produce fringes of equally high contrast, giving rise to high percentage modulation of the beam. These Fresnel images repeat at intervals of Np^2/λ where N is an integer. However, the fringe contrast is still subject to the effects of imperfect collimation, and only the first image is of practical importance.

The first Fresnel gap is plotted against grating pitch in Fig. 7 and it will be seen that it enables the use of 100 lines mm^{-1} gratings at a gap of 0·1 mm. For 50 lines mm^{-1}, however, the Fresnel gap of 0·44 mm is appreciably greater than the maximum collimation gap of 0·24 mm. To overcome this, a phase grating is used for the first grating, usually the index, in conjunction with a slit and bar scale grating. This has the effect of moving the family of Fresnel images by one half Fresnel interval of $p^2/2\lambda$ so that there is now zero contrast with the gratings in contact and maximum contrast at this half interval which is plotted in Fig. 7.

This arrangement gives an optimum working gap of 0·22 mm for 50 lines mm^{-1} gratings, which is clear of the 0·24 mm maximum collimation limit. A tolerance of $\pm 25\%$ of the nominal gap gives no more than 10% reduction in modulation.

The requirement for the phase grating is that the bars should have an optical retardation of approximately $\lambda/2$. The exact shape of the contours is not critical and any shape from a serrasoidal to a square castellation is acceptable. An approximately sinusoidal contour is achieved by controlled bleaching of a slit and bar image in a photographic emulsion whereby swelling of the gelatin in the region of the silver image produces the required surface corrugations. This arrangement has an additional advantage in that it is more efficient than a slit and bar grating pair, since no light is absorbed by the first grating.

The above considerations assume the use of monochromatic light. In practical systems there is some spread of effective wavelength which results in a further reduction in contrast of the fringe but which has the advantage of largely eliminating harmonics which are present to an appreciable extent in the Fresnel image.

3.3.3. Spectroscopic systems

Prior to the availability of economic electronic methods of pitch subdivision by greater than a factor of four, it was necessary to use gratings in the range 100–250 lines mm^{-1} to achieve resolutions of 1 or 2 μm.

If both index and scale gratings are of the phase variety arranged to give approximately equal energy in the zero and first order spectra, and if an output slit is positioned so as to receive light from the grating pair in the direction of the first order spectrum only, good modulation is obtained which is limited only by collimation effects.

This gives a working gap of 0.1 mm with 125 lines mm^{-1} gratings and 0.05 mm at 250 lines mm^{-1}. The smaller gap is acceptable on small-scale applications such as the Moore & Wright digital micrometer which uses this arrangement in order to minimise electronic complexity in a portable instrument. The gratings used are chrome on glass slit and bar type which are more robust than the phase variety and which give equally good fringe contrast at the expense of some loss of light energy.

3.3.4. Three grating systems

A system has been developed at the National Engineering Laboratory (NEL)[9] which uses a set of three gratings illuminated with relatively diffused light. The first and third gratings are fixed relative to each other and critically spaced from the second which moves relative to them.

The theory of operation is rather complex but basically the second grating, acting as a form of linear Fresnel lens, projects an image of the first grating in the plane of the third grating.

It is claimed that with suitable gratings of fine line structure, fringes of high contrast can be obtained, at much greater working gaps than with two grating systems, without the need for a high degree of collimation.

4. DISPLACEMENT MEASUREMENT

4.1. Grating Replication Methods

The earliest metrological gratings were produced at the National Physical Laboratory (NPL) using the Merton process.[10] These were cast resin replicas on glass of 2500 lines in^{-1} and operated with a spectroscopic type reading head.

They were used successfully on a number of scientific applications such as star plate measurement, analysis of bubble chamber stereograms and on stereo comparators in photogrammetry. However, the multiple section process did not lend itself to large scale production.

A photographic copying method which allowed for in-process error correction was developed at NPL and further improved at NEL in the 1960s. Known as NPL Method III,[11] this gave high accuracy gratings on glass of line structures from 100 down to a few lines mm^{-1}. These were widely used on coordinate measuring machines and on small machine tools for digital readout.

Metrological gratings were first applied to numerically controlled machine tools by Ferranti in 1954.[12]

A group at Staveley Research developed an electronic method of pitch subdivision which enabled relatively coarse line structures to be used.[13] These were particularly applicable to large machine tools with travel lengths up to 10 m. Multi-section glass assemblies gave rise to problems on installation and to meet this need, a method of etching a scale on to long lengths of flexible stainless steel tape was developed by Technicolor Ltd, in collaboration with NPL. At the same time Ferranti Ltd developed stainless steel gratings on a rigid base printed from masters generated by the NPL Method III process. (Ferranti have since taken over the Technicolor plant and process.)

There is a need for both types of steel grating which are read by reflection. The rigid type is preferable for fine line structures and short travel lengths. The flexible type is used for larger scale installations and for wraparound applications.

A range of glass and steel gratings are made by Heidenhain in West Germany. Bausch & Lomb in the USA produce chrome on glass gratings. OMT in the UK produce radial and linear glass gratings using the NPL Method III process.

4.2. Photoelectric Fringe Detection

The basic elements of a displacement measuring system are shown in Fig. 6, consisting of light source, collimating lens, index grating, scale grating and photosensitive detectors. A single detector will indicate extent of movement but not direction. A minimum of two detectors spaced preferably one quarter fringe pitch apart give rise to output waveforms in quadrature. These represent a lead or lag of electrical phase which may be interpreted by suitable circuitry to determine the direction of movement and hence to direct pulses representing positive or negative increments of movement into add or subtract channels of subsequent counting circuitry.

In practice four photodetectors are normally used[14] arranged as shown in Fig. 8. These derive signals from four areas of the fringe spaced at quarter pitch intervals. The alternative counter phase pairs of signals from areas 1, 3 and 2, 4 are combined in push-pull amplifier stages so that the fundamental modulations add while d.c. levels and even harmonics cancel

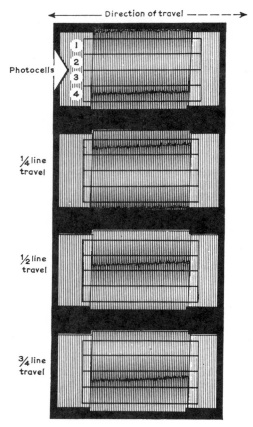

Fig. 8. Four-phase fringe detection.

resulting in a pair of quadrature signals symmetrical about a mean d.c. level (Fig. 9).

Where space is restricted, a d.c. compensating signal may be taken directly from the light source on a single photodetector and used to balance both modulated channels.

The reading head must provide a beam of collimated light to cover the area of the fringe. This can be produced by a small tungsten filament lamp with the filament parallel to the grating lines and a simple aspheric lens. The four photodetectors consist of silicon photodiodes each with an area of $10 \, mm^2$. Since the photodetectors are infrared sensitive the lamp may be

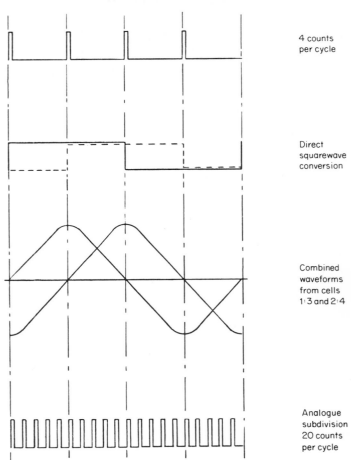

4 counts
per cycle

Direct
squarewave
conversion

Combined
waveforms
from cells
1:3 and 2:4

Analogue
subdivision
20 counts
per cycle

Fig. 9. Two-phase combined signals from four phase detectors.

run below nominal voltage for a long life. A solid state GaAs light-emitting diode source is a possible alternative. This type is well matched to silicon detectors but it operates at a lower power level than even a small filament lamp. Care must be taken to maintain the signal level in the reading head well above the highest level of electrical interference likely to be encountered in a machine environment. For this reason phototransistor detectors which have intrinsic gain are used in conjunction with a solid state source, although their small surface area receives light from a limited area of grating unless auxiliary optics are used.

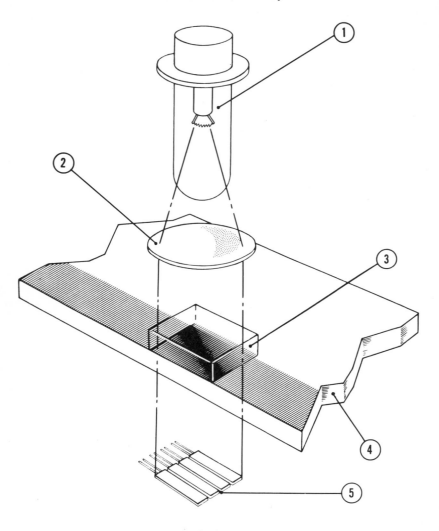

KEY

1	Exciter lamp	4	Scale grating
2	Collimating lens	5	Photocell strips
3	Index grating		

Fig. 10.   Optical arrangement used with slit and bar transmission gratings.

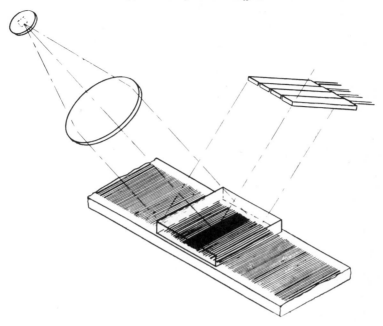

Fig. 11. Optical arrangement used with reflection gratings.

Figure 10 shows the optical arrangement in a reading head used with transmission gratings of identical pitch.

Figure 11 shows the arrangement with a reflecting scale grating. In this case, to suit the reflective geometry, a vernier fringe is used. This runs parallel to the grating lines and uses an index grating differing by one line over the field width and aligned parallel to the scale grating.

4.3. Self-Contained Linear Transducers

To allow for easy installation on the general run of machine tools self-tracking transducers have been introduced. These are of two main types, the push-rod and the side-linked, both of which have an integral slideway and carriage on which the reading head assembly is preset. A small amount of lateral flexibility allows for coupling to the machine members without affecting this setting. Because of the relatively short base of the carriage, grating pitches are preferably restricted to the range 10–20 lines mm^{-1}.

The push-rod type has an annular seal around the rod which is extremely

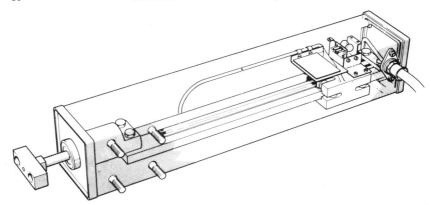

Fig. 12. Push rod linear transducer using steel reflection grating. (Courtesy of Ferranti Ltd.)

effective in adverse environments. It is suitable for travel lengths up to about $\frac{1}{2}$ m and is particularly applicable to the in-feed axis on lathes and borers (Fig. 12).

The side-linked variety has a sliding seal and utilises an extended enclosure. Lengths up to 3 m can be achieved, and this type is used on milling machines and on the longitudinal axis of lathes (Fig. 13).

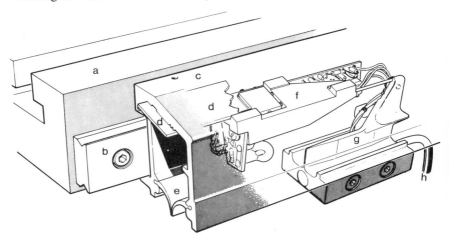

Fig. 13. Self-tracking linear transducer using steel reflecting grating. (a) Machine table; (b) mounting bar; (c) U-section extruded housing; (d) scale grating; (e) seals; (f) reading head; (g) drive block; (h) cable to associated display unit. (Courtesy of Ferranti Ltd.)

Small rotary transducers incorporating radial gratings may be used for linear measurement by coupling via a friction wheel, rack or leadscrew, although these cannot be regarded as giving direct linear measurement.

4.4. An Afocal Projection Reading Head

To cater for linear measurement or machine travels outside the capacity of self-contained transducers it is necessary to revert to an arrangement in which the reading head is a separate unit attached to one member with the scale grating mounted on the other. The desirable features of such a reading head can be summarised as follows:

1. The working gap should be in the range 5–10 mm with a tolerance of at least ±0·5 mm.
2. Tolerances on mounting squareness should be not less than ±5 milliradians.
3. The working field should not be less than 1 cm² so as to give continuity of signal in the presence of swarf particles and to be capable of reading over joins between grating sections.
4. A working signal should be maintained over moderate levels of contamination of the grating with oil, swarf, coolant, etc.
5. The associated electronic display unit should operate over a wide range of signal amplitudes and should provide early warning and system failure indication at successively reducing signal levels.

To meet these requirements an afocal projection arrangement is appropriate in which light sources, reflector and photodetectors are arranged in one focal plane of a set of lenses while the scale grating is in the other focal plane.

To maintain d.c. balanced signals with a partially obscured field requires the use of four separate optical channels each deriving a signal from the same area of grating but introducing a spatial phase shift of 90° between adjacent channels and being directed on to a separate photodetector.

A compact light source at the upper focus of lens 1 produces a collimated light beam which intercepts an area of grating at the lower focus of lenses 1 and 2. An image of the source is produced at the upper focus of lenses 2 and 3 and is reflected by a mirror so as to produce a collimated beam from lens 3. This projects a reversed image of the grating field on to an adjacent area of the grating. If the grating moves in one direction the image moves an equal amount in the opposite direction and modulation is produced at the rate of two cycles per grating pitch. This enables a coarser grating to be used

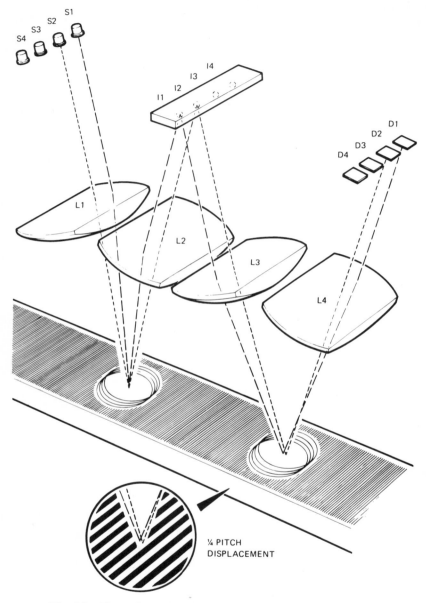

Fig. 14.   Four channel optical paths in afocal reading head.

for a given system resolution than that required with a scale and index arrangement. For instance a 2·5 lines mm^{-1} grating gives a cycle length of 0·2 mm which would normally be produced from a 5 lines mm^{-1} grating. Fringe subdivision in the range 20–100 gives digit sizes in the range 10 to 2 μm. The four modulated final images are accurately aligned on photodetectors during assembly. The optical arrangement ensures that this alignment is relatively insensitive to small changes in position between reading head and grating.

If the energy from the whole area of the overlapping fields were used to produce signals there would be a possibility of introducing imbalance if the edge of one field was obscured while not affecting the antiphase field. To avoid this it is arranged, by means of a combined mask and splash guard, that the photodetectors receive energy only from the central area which is common to all four channels.

Figure 14 shows the complete four channel optical arrangement with corresponding central rays of channels 1 and 2 traced through the system.[15]

Figure 15 shows the head associated with the grating mounted in an extruded spar on a machine tool slideway.

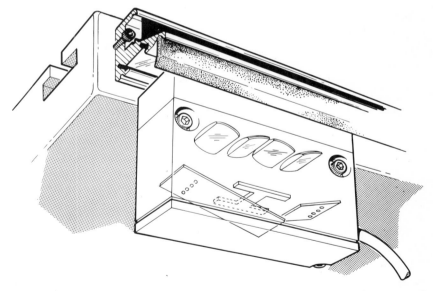

Fig. 15. Afocal projection head and grating spar on machine slide. (Courtesy of Ferranti Ltd.)

4.5. Electronic Signal Processing

The electronic part of the system, located at some distance from the measuring transducers, has to receive the amplified photodetector signals and convert them to bidirectional count pulses. These are added or subtracted on a multistage decimal counter and displayed as a digital readout on illuminated display devices.

Seven-bar type displays incorporating light-emitting diodes (LEDs) have mostly replaced the neon gas discharge type.

There are two main different forms of transducer output signal to which most transducers conform (Fig. 9). The simpler transducers provide two interlaced square waveforms at transistor–transistor logic (TTL) level, i.e. in the range 0 to $+5$ V. These are at grating pitch frequency and give rise to digits at the rate of four per grating pitch.

A common arrangement uses 25 lines mm^{-1} gratings giving a digit size of $10\,\mu$m. This type of signal cannot be further subdivided.

For finer subdivision the serrasoidal form of the signals derived from the fringe must be maintained. These are interfaced as a pair of signals symmetrical about zero and usually 10–20 V in amplitude. They may be subdivided into between 10 and 100 digits giving resolution down to $1\,\mu$m from 10 lines mm^{-1} gratings. This arrangement gives less critical tolerances in the reading head than with 25 lines mm^{-1} gratings. The frequency bandwidth is lower and this, together with signal filtering at the interface, gives considerably better signal-to-noise discrimination.

These analogue signals may be monitored for reduced amplitude so that early warning and system failure indication can be given at successively reducing amplitude.

4.6. Accuracy of Moiré Fringe Transducers

The accuracy of the transducer is mainly determined by the line placement accuracy of the scale grating. It is limited by the digital resolution of the system which depends on the grating pitch and the degree of subdivision of the pitch cycle.

On complete machine installations the largest component is quite often the Abbé error due to the distance between the scale and the point of measurement. This error has a magnitude of $1\,\mu$m per 200 mm offset for each arc-second of angular change due to curvature or slackness of slideways. It cannot be attributed to any fault in the transducer but is a frequent source of complaint when installed systems fail to meet the specification relating to the transducer alone. It can be minimised by mounting the scale grating or transducer as close as possible to the line of measurement.

The transducer accuracy specification can relate only to displacement between the reading head assembly and scale grating. It does not usually include an allowance for digital resolution which is determined by electronic circuitry elsewhere in the system. An allowance of one digit should be made for the uncertainty of positioning due to the digital nature of the system. Also, where a high degree of pitch subdivision is used, an allowance of between one hundredth and one thirtieth of a pitch should be made for analogue subdivision of the cycle assuming d.c. shift in the range 3–10 % of the peak to peak amplitude.

The digital resolution and pitch subdivision errors affect short range accuracy but are negligible over long distances compared with other long range errors. Scale errors over distances comparable with the reading head field width of around 10 mm are extremely small, of the order of 1 μm, due to the integrating effect of the moiré fringe. Error peaks of a few micrometres can occur over distances between 10 and 100 mm and these can increase to 10–20 μm over 1 m travel lengths. Long range errors can be controlled relatively easily during grating replication and can be reduced to zero between any two specified points by the use of a cosine law correction technique. This is achieved by printing the grating lines at a slight angle of the order of 20 milliradians to the scale axis. An adjustment of the scale axis relative to the line of measurement introduces correction at the rate of $\pm 20 \mu$m for ± 1 milliradians of skew.

A single figure accuracy specification does not differentiate between long range and short range errors. An expression of the form $e = \pm(a + bL)$, where a represents short range error and b is error per unit length L, is unduly pessimistic over long distances. An alternative formula $e = \pm(a + b\sqrt{L})$ is a useful practical compromise.

The values of a can be 1–5 μm and $b\sqrt{L}$ over 1 m can be 5–20 μm. Higher accuracies can be achieved at great cost but are of doubtful practical significance relative to thermal expansion which for steel is 11 μm m^{-1} °C^{-1}.

4.7. Angular Measurement

The moiré effect operates on radial gratings printed on glass discs provided a suitable matching index piece is used. Slight curvature of the fringes does not adversely affect the operation.

Relatively coarse line structures of a few hundred lines are used on small discs for rotary transducers suitable for use on racks or leadscrews to give linear measurement.

For direct angular measurement discs are usually 250 mm in diameter with line structures of 9000–36 000 giving angular pitch of 0·04° to 0·01°. This can be subdivided to digital resolutions of 0·001° to 0·0001° with overall accuracies of the order of ±0·001°.

These discs can be mounted in rotary tables concentric with high quality bearings to give angular measurement comparable with the accuracy of the grating. For extreme accuracy a pair of reading heads may be used at each end of a diameter and the signals combined so as to cancel errors due to eccentricity. The signals may only be added directly provided eccentricity does not exceed about one-eighth pitch otherwise severe amplitude reduction may occur. Alternatively, each channel may be separately digitised and the resultant digital signals combined.

Where very large cylindrical surfaces are available in the range 1–10 m diameter, flexible stainless steel grating tape can be wrapped round these surfaces slightly tensioned and clamped at intervals. Lengths of tape over 4 m can be successfully jointed *in situ* to give effectively a continuous scale. Provided the total number of lines is known measurement can be converted into a degree scale using a microprocessor-based ratio converter.

This grating arrangement can be used either with a conventional or afocal projection type of reading head provided due allowance is made for the effect of curvature on the optics.

5. DISPLACEMENT MEASURING APPLICATIONS

5.1. Machine Tool Slide Position Measurement

Most installations on small to medium size machine tools use the self-tracking type of linear transducer which has been preset for optimum signal quality. The transducer is attached to the edge of the machine table while the drive block with some lateral compliance to take up tolerances is attached to the body of the machine.

The fully sealed pushrod transducer is particularly suitable for application to the in feed axis on lathes and borers.

For large machines of greater than 3 m travel the afocal projection reading head can be installed with relatively uncritical tolerances in association with a grating held in an extruded spar.

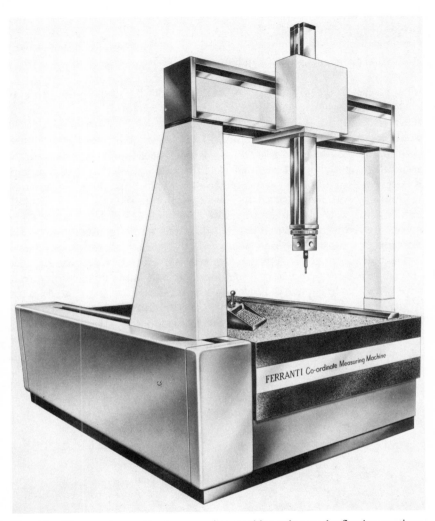

Fig. 16.   Three-axis coordinate measuring machine using steel reflection gratings.
(Courtesy of Ferranti Ltd.)

For medium to large machines which do not require the accuracy of a true linear scale an economic installation can be achieved using a friction driven rotary transducer.

5.2. Coordinate Inspection Equipment

The moiré fringe technique is ideally suited to the requirements of coordinate measuring machines. These require operation to an accuracy at least half an order of magnitude better than the tolerances on the components they are inspecting. They have high precision linear movements so that scale gratings, usually of 50 lines mm^{-1}, can be mounted directly on one member with the reading head on the other with an accurately maintained working gap of 0·2 mm. Long range error can be minimised by applying the cosine law correction to the grating *in situ*. This can be done so that in the presence of slight residual slide curvature Abbé errors are reduced to zero in the middle of the working area although the gratings are mounted to one side.

Since the original two axis machine was introduced by Ferranti in 1959, the range has been extended to include three and four axis machines up to several metres in capacity. Power drives and data processing facilities are available as options with full computer numerical control at the top of the range. An example of a Ferranti three axis machine is shown in Fig. 16. An

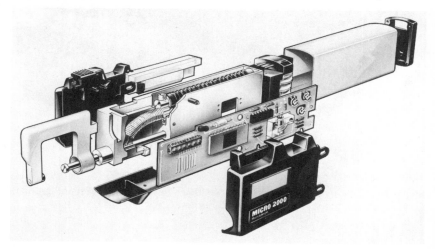

Fig. 17. Digital micrometer using glass transmission grating. (Courtesy of Neill Tools Ltd.)

interesting small scale application of moiré fringes is in the Moore & Wright digital micrometer. This uses a 250 lines mm^{-1} glass grating to achieve a 1 μm digit size with divide by four pitch subdivision. A complex custom designed microcircuit counts and displays position on miniature seven bar readouts. The unit is shown in Fig. 17.

6. APPLICATIONS IN EXPERIMENTAL STRESS ANALYSIS

6.1. Principles of Strain Measurement

When the members of a structure are stressed the members are subjected to distortions which can in general be subdivided into components parallel and perpendicular to the surface of the members. This subdivision arises because the principal components of strain lie in these directions (providing no external forces are applied at the location of interest). The main types of load induced distortion can be summarised as bending, shear, torsion and direct stress. The first three types produce significant out-of-plane displacements of the member, whilst the latter produces very small out-of-plane displacements, caused by Poisson contraction of the member. In the most general case, all four types of distortion can coexist, and the in-plane and out-of-plane distortions must be determined at every point in order to analyse the stress distribution completely. It is more usual, however, for one type of loading to dominate and the method of analysis is selected to determine the effects of this loading.

In experimental stress analysis, the stress distribution is determined from measurements on the distortion of either models of the structure, or on actual structures. In this context, the moiré effect can be used in five ways:

(i)   extensometers;
(ii)   measurement of normal deflections over large surface areas;
(iii)   measuring flexure over large surface areas;
(iv)   measuring in-plane displacement over large surface areas;
(v)   measuring in-plane displacement of structural models.

Apart from the application to extensometers, the moiré effect is utilised as a large field technique, capable of measuring variations in strain and displacement within the area of any one grating.

6.2. Extensometers

Extensometers are mechanical instruments for measuring strain over a fixed gauge length. The displacement between two points is picked up by a lever system, and either read directly after further mechanical magnification or converted to either an optical or electrical signal by use of a subsidiary transducer. The moiré effect has been used in a number of different instruments based on this principle[16] and a typical example is illustrated in Fig. 18.[17] The frame of the instrument carries a fixed grating and the lever system, consisting of a 10:1 driving lever plus an idler arm, carries the displacement grating. Gratings of 100 and 99 lines in^{-1} with mark/space ratio of 6·25:1, give sharpened fringes every inch, and an overall magnification of 1000:1. On a gauge length of 25 mm, strains of 10^{-5} could be detected by visual observation, and electronic subdivision could increase this accuracy.[18] Other instruments were developed commercially, but the writers are not aware of any such extensometers currently on the market.

The main advantage of these extensometers is their high sensitivity combined with a large range, a feature of the moiré effect. They are also capable of operating at very high temperatures ($>400\,°C$), a notorious problem in accurate extensometry. Their main drawback is their high cost, which makes them non-competitive at ambient temperatures with inductance transducers and similar instruments. The latter perform the same function, although with less accuracy for large ranges, at a much smaller cost.

6.3. Determination of Out-of-Plane Deflection

The determination of surface deflections is required in problems on plates and shells, where bending stresses are the most important. Such measurements are especially useful where instability is the likely cause of failure. The actual determination of bending moments from deflection data alone is unsatisfactory as the curvature must be calculated. This involves double differentiation of the deflection data, which reduces the accuracy considerably.

Deflection measurements are also used in plane stress problems to measure the third principal strain, ε_z, and hence to separate the principal stresses in photoelastic investigations.

Two methods are available for measuring deflection by moiré.

6.3.1. Shadow technique

If a reference grid, placed close to a diffuse reflecting surface, is illuminated with a collimated light beam inclined to its surface then moiré fringes will be observed between the grid and its shadow, when viewed normally. The

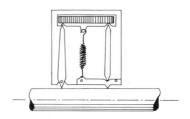

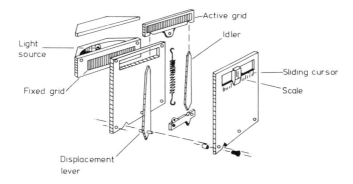

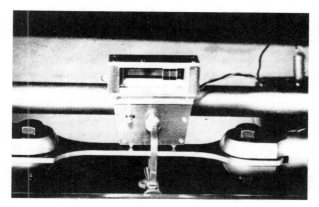

Fig. 18. Basic layout of the moiré extensometer.

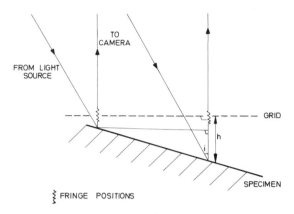

Fig. 19.   Principle of moiré shadow method.

fringes are due to the angular difference between illumination and observation lines, and to the variation of distance between specimen surface and reference grid.

The relation connecting these quantities is (Fig. 19):

$$h = \frac{p}{\tan i}$$

where h is the contour interval, as the fringes represent contours of constant deflection. Taking $p = 0 \cdot 1$ mm and $i = 20°$, contours occur every $0 \cdot 25$ mm. The initial separation of specimen and grid, and the value of i must be adjusted to give fringes of maximum contrast.

Experimentally, this technique is simple and the fringe patterns are easy to analyse. If a specimen with a slightly curved surface is used, however, a coarser grid (with 10 or 20 lines mm^{-1}) will give much greater depth of field, but with corresponding decrease in sensitivity. This also applies if deflections exceed about $0 \cdot 25$ mm (4 or 5 fringes).

The main advantage is that fringes can be observed on normal machine or cast finish, provided any oxide is removed. Its disadvantage is the proximity of the grid to the specimen, usually less than $2 \cdot 5$ mm.

6.3.2. Projection method

The same geometrical arrangement can be achieved by projecting a grid onto the specimen and viewing the grid through a similar optical system with the master grid placed in its image plane, as in Fig. 20. Alternatively,

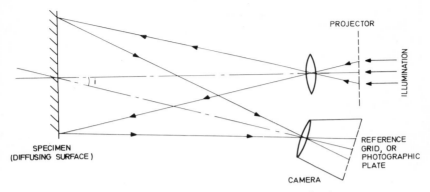

Fig. 20.   Projection method for determination of deflection contours.

the projected grid may be photographed before and after loading by the camera and so any initial deformation in the specimen surface cancelled out.

It is important to keep both projector and camera at similar distances from the specimen, due to the perspective effect (the change of magnification caused by change in object distance) of their lenses. Identical object distances cancel out this effect. It is also advantageous to use identical lenses so that the lens distortion will also cancel out.

The equation connecting contour interval with angle of incidence is the same as before. The pitch is now that of the projected image, and use of conventional projectors will limit this to a maximum of about 10 lines mm^{-1}. This is produced by enlarging a 40 lines mm^{-1} grid, in the normal slide position of the projector by 4 ×. The sensitivity is not as great as the shadow method due to this limitation, but this technique may have applications for use on heated surfaces. The method worked quite successfully on red hot steel plates, yielding contours at approximately 0·25 mm intervals over the plate. More detailed theory and applications are given in a series of papers in ref. 39.

6.3.3. Applications
Both techniques for measuring out-of-plane deflections are simple to apply, and the fringe patterns can be easily interpreted. The technique is most useful where the shape of the fringe pattern can give useful information, rather than just a quantitative assessment of deflections. This aspect has been used by Luxmoore to demonstrate the antilastic curvature of beams to students (and also measure the Poisson's ratio). The techniques have also

been used to observe and measure the buckling behaviour in deep shear webs, covering areas of $1 \, m^2$.[19] This problem is particularly difficult if measurements are only made at a small number of locations, using individual transducers.

The techniques can also be used to study vibration patterns, in a manner analogous to time-lapsed holography (Chapter 6). On vibrating surfaces, antinodal positions will give continuously varying deflections, whilst nodes will produce zero deflection. A time-lapsed photograph of a grating projected onto a vibratory surface will produce areas of washed out grating at the antinodes whilst the nodal positions have a sharp grating image. If a grating is superimposed on the photograph to produce dark fringes at the nodes, the antinodes will appear light, and one immediately has a contour map of the nodal positions.

6.4. Measurement of Slope

The experimental determination of slope is most important in models of plate and shell structures. Two methods, both very similar, have been devised for this purpose. Both depend on a combination of the optical lever and the moiré effect, and they are an example of an experimental set-up of form 3.

In the Ligtenburg apparatus,[20] a screen is viewed via the reflecting surface of a plate model as in Fig. 21. The grid, which is paper with only

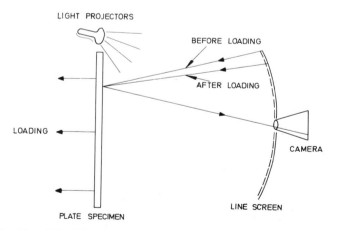

Fig. 21. Principle of Ligtenburg's apparatus for measuring slope.

1 line mm^{-1} printed on it, is slightly curved to correct for the perspective effect of the camera. Usually a photograph of the grid is taken before loading to allow for any initial distortion of the model in the reference grid. When the model is loaded, the camera sees a distorted view of the paper grid, which is then photographed again. The two photographs are then viewed together or exposed on the same film. In the original apparatus each fringe, which represented a contour of constant slope, constituted a change in angle of 0·001 radians.

In Theocaris's modification[21] of the above system (Fig. 22) a telecentric

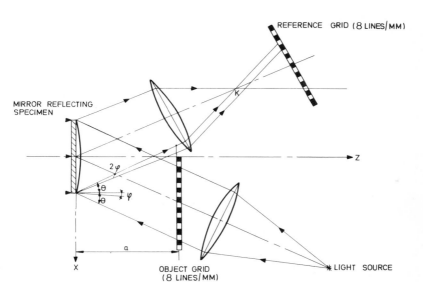

Fig. 22. Apparatus due to Theocaris for determining slope contours.

lens system is used to remove the perspective effect of a conventional camera. A higher sensitivity is obtained—about 10^{-4} radians.

As the fringe patterns represent contours of constant slope, it is quite easy to determine the components of curvature from them. At any point on the specimen, the slope may be specified by two components $\partial w/\partial x$, $\partial w/\partial y$ measured along rectangular coordinates (x, y). To obtain this information, it is necessary to align the target screen along both the x- and the y-axes, so that first the $\partial w/\partial x$ contours, then the $\partial w/\partial y$ may be recorded. Graphical differentiation along the x- and y-axes yields $\partial^2 w/\partial x^2$ and $d^2 w/\partial x\,\partial y$ from the $\partial w/\partial x$ contours. Applying the same procedure to other contours gives $\partial w^2/\partial y^2$ and $\partial^2 w/\partial y \, \partial x$. Because of compatibility

$$\frac{\partial^2 w}{\partial x \, \partial y} = \frac{\partial^2 w}{\partial y \, \partial x}$$

so that the bending moments may be calculated from

$$M_x = -D\left(\frac{\partial^2 w}{\partial x^2} + \frac{v\partial^2 w}{\partial y^2}\right)$$

$$M_y = -D\left(\frac{\partial^2 w}{\partial y^2} + \frac{v\partial^2 w}{\partial x^2}\right)$$

$$M_{xy} = M_{yx} = -D(1-v)\frac{\partial^2 w}{\partial x \, \partial y}$$

where v is Poisson's ratio and D is the flexural rigidity.

It is quite feasible to design portable apparatus based on the above designs, for use on models in difficult locations. This has been done using the Ligtenburg principle.

6.5. Measurement of Surface Strain

6.5.1. Basic principles

For in-plane strain measurement, a grating is deposited on to the specimen surface, either by bonding, as with conventional strain gauges, or by photoengraving. A transparent reference (or master) grating is super-imposed either in contact or in the image plane of a projection system. As the specimen is strained, moiré fringes will be formed due to changes of pitch and angle in the specimen grating. It is easily shown that the fringes represent contours of in-plane displacement, as measured perpendicularly to the grating lines. For in-plane deformation, two orthogonal displacement parameters, usually designated u and v, are necessary, and the strain parameters are related to these by:

$$\varepsilon_x = \partial u/\partial x \qquad \varepsilon_y = \partial v/\partial y \qquad \gamma_{xy} = \partial u/\partial y + \partial u/\partial x$$

Hence two orthogonal gratings are needed on the specimen for a complete strain evaluation, and these may be deposited simultaneously, producing an orthogonal array of dots (Fig. 23). To avoid confusion between the two fringe patterns (referred to as u- and v-isothetics), the master grating remains a line grating, aligned first with one specimen grating, then rotated

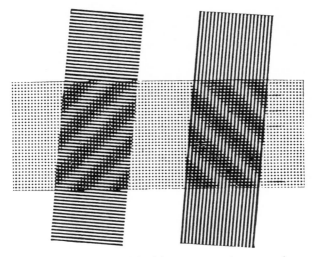

Fig. 23. Orthogonal specimen grid with master gratings superimposed at right angle.

through exactly 90° to be aligned with the other grating. This procedure ignores the specimen grating currently at 90° to the master grating.

From Fig. 1 and eqn (1), it can be seen that the normal strain ε_x (or ε_y) can be obtained by measuring fringe spacings perpendicular to the grating lines, and the component of shear $\partial u/\partial y$ (or $\partial v/\partial x$) by measuring parallel to the grating lines. For a varying strain field, it is usual to plot the fringe order (equal to some multiple of the grating pitch) against distance along an orthogonal network of lines parallel and perpendicular to the grating lines (Fig. 24) and determine the derivatives graphically at the points of intersection of the network. This is done on identical networks for both specimen gratings and then the complete strain information is obtained at each intersection of the network.

6.5.2. Sensitivity
The main disadvantage of the conventional moiré technique outlined above is lack of sensitivity. Gratings are available commercially for moiré strain measurement with densities of up to 40 lines mm^{-1} (1000 lines in^{-1}). Strains can be measured from direct fringe patterns using these gratings with a sensitivity of around 1000 microstrain, which is outside the working strain range of many common structural materials. Suitable interpolation techniques can increase this sensitivity by an order of magnitude.

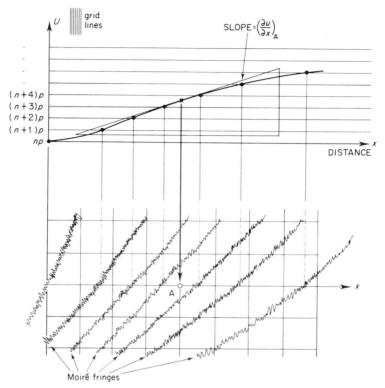

Fig. 24. Analysis of strain at points of intersection of network using displacement–distance plot.

The most common interpolation procedure is the initial mismatch between master and specimen grating. This can be an angular misalignment, a difference in pitch, or a combination of the two. The mismatch is constant over the grating area, producing parallel fringes of equal separation. Changes in the specimen grating will distort the initial fringe pattern, and analysis can be carried out by subtracting the apparent initial displacements from the final displacements.

This procedure also simplifies the determination of tensile and compressive strain, for if an initial pattern is produced by making the master pitch larger than the specimen pitch, a tensile strain will increase the fringe spacing, and vice versa for a compression. In practice, where the sign of the strain is known in advance, it is always best to arrange the mismatch so that there is an increase in the number of fringes.

With a mismatch, the sensitivity depends on the accuracy with which the fringe centres can be located. Experience has shown that, even with electro-optical methods of fringe location, this is at best accurate to 1 % of the fringe spacing. Hence the displacement sensitivity is no more than 1 % of the pitch, and the strain sensitivity will depend on the effective gauge length over which the displacement is measured. If this is large, e.g. a constant strain field over the area of the specimen grid, then the strain sensitivity is usually well within the elastic range (< 0.1 %) of most materials.

A mismatch between master and specimen grating extends the range of the moiré technique by producing fringe patterns in strain fields where no fringes would occur with the direct moiré effect. This is very useful for constant strain fields, but does not help with plotting strain gradients. In fact, a large mismatch will often obscure a small strain gradient: the mismatch will produce closely spaced fringes in which small variations of fringe spacings will be ignored.

The plotting of strain gradients can be improved by including the positions of fractional fringe orders (cf. photoelasticity). These can be obtained by moving the master grating a fraction of a pitch perpendicular to the grating lines. If, for example, this fraction is one half pitch, then the fringes produced correspond to displacements of $(n + \frac{1}{2})p$, $(n + 1\frac{1}{2})p$, $(n + 2\frac{1}{2})p$, etc., as shown in Fig. 25. The subdivision may be finer than this,

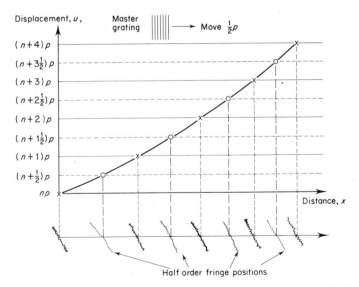

Fig. 25. Plotting half order fringes by moving master grating one half pitch.

of course, but there is no point in plotting fractions smaller than 0·1 pitch, as the uncertainty in determining the fringe centre will approach this value. Also movement of the master grating must be very precise.

This procedure for plotting fractional fringe orders does not increase the basic sensitivity of the moiré technique, but the mismatch technique does; thus these two interpolation methods are complementary rather than alternatives. They are essential for elastic strain evaluations.

Attempts to improve moiré sensitivity have been aimed at increasing the line density of the specimen grating. Line densities greater than 100 lines mm^{-1} are not generally available for moiré work because of the difficulty of reproducing these gratings photographically, and even when they are available, formation of moiré fringes requires special spectroscopic systems.[22,23] The sensitivity of existing gratings can be improved by utilising their diffraction effects or using improved interpolation methods.[24] Alternatively, very fine gratings (up to 3000 lines mm^{-1}) can be produced interferometrically, with the help of a laser.[25,26]

6.5.3. Some practical techniques

Suitable gratings ranging from 2 up to 40 lines mm^{-1} can be obtained commercially,[27] including gratings on photographic stripping film which can be bonded directly to components. Thin metal gratings can also be obtained,[28] and these can be bonded directly to components, or used as stencils for printing gratings onto a surface.

The simplest method of observing the moiré fringes is to superimpose the master grating directly onto the specimen grating. For unidirectional strain measurement, this can be done by using a master grating on film, and adhering it to the specimen grating with a clear grease. This allows the specimen to distort freely, without distorting the master grating, but compression of the specimen can cause the film to buckle if it is not stiffened in some way. Where a complete two-dimensional strain analysis is required, the master grating must be capable of rotation through 90°, and it may prove more convenient to either photograph or replicate the specimen grating. Details of these techniques and other applications are given in ref. 24.

6.5.4. Displacements of structural models[29]

A modification of the previous technique has been applied to two-dimensional model analysis. A model made of a suitably elastic material is printed overall with a fairly coarse grating (4 lines mm^{-1}). The complete model is then viewed through a corresponding reference grating and loaded. The coarse grating is insensitive to the strains, which are small, but sensitive to the rigid body rotations, which are large. This enables the flexure of the model components to be measured, and hence the influence lines may be determined.

6.6. Automatic Fringe Analysis

Most of the automatic fringe analysing systems have been applied to in-plane strain measurements, although they can be applied just as well to the other applications of the moiré effect in stress analysis. They all depend on the optical filtering inherent in most optical systems (see Section 3) which reduces the intensity variation across the fringes to a sinewave distribution, expressed as[30]

$$I(x) = I_0 + I_1 \cos \left(\frac{2\pi}{p} \rho(x) \right)$$

where $\rho(x)$ is the relative displacement between the two gratings, i.e. the function that is required. Hence

$$\rho(x) = \frac{p}{2\pi} \arccos \left(\frac{I(x) - I_0}{I_1} \right)$$

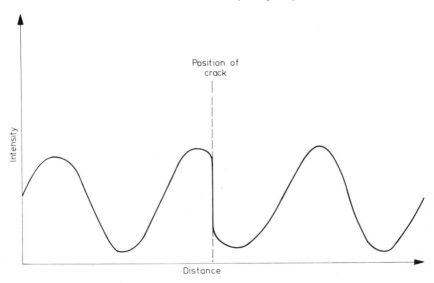

Fig. 26. Discontinuity in vernier moiré fringe intensity trace due to crack in structure.

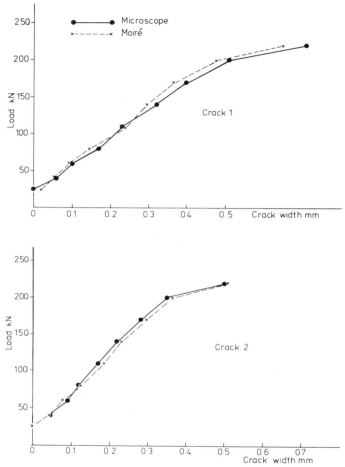

Fig. 27.   Comparison of microscope and moiré measurements of crack widths in a reinforced concrete beam.

Unfortunately, this relatively simple expression is complicated in practice by the presence of noise and random variations in I_0 and I_1, which are not easily calibrated. Sciammarella[30,31] has tackled this problem with a combination of analogue and digital filters, after obtaining the intensity distribution from a microdensitometer trace of a photograph of the fringes. Luxmoore[32,33] has avoided the variations in I_0 and I_1 by differentiating the signal to obtain the positions of the peaks and troughs, but this accentuates the effect of the noise. Recently, he has used digitised data from a television camera with a least squares numerical technique to fit a sinewave to the data over a distance of at least one fringe. This provides a suitable averaging procedure which reduces the effect of the noise, and gives the required data in the form of average frequency and phase over the region fitted. This technique has been applied to the study of cracking in concrete, using gratings of 2 lines mm^{-1} bonded to reinforced concrete beams. The cracks caused discontinuities in the fringe pattern, with a corresponding effect on the intensity trace (Fig. 26). By fitting sinewaves to traces on either side of the cracks, the difference in phase could be determined, and multiplying this phase difference by the pitch gave the crack width. The method was compared with microscope measurements, and very satisfactory results obtained (Fig. 27).

7. MISCELLANEOUS APPLICATIONS

The moiré effect, combined with the different optical systems described in Section 1, is capable of a wide range of applications, and some of the less well known examples will be discussed here.

Pearce[34] has used the shadow moiré technique to monitor the shape of quasi-hemispherical shells. A narrow grating of 20 lines mm^{-1} is curved along its length to provide a small gap between itself and the shell, which is mounted on a lathe whose axis is parallel to the grating. The fringes are observed either photographically, by means of a drum camera which drives the film synchronously with the lathe, or with a scanning photomultiplier. Rotation of the shell enables a complete contour map to be obtained. It is intended that a master shell is placed in position, and a datum map obtained. Subsequent shells can then be turned on the lathe until the desired shape is obtained, using the moiré pattern to indicate any deviation between the two shapes. The technique is sensitive to displacements of 2·5 μm.

An interesting automatic instrument also based on the shadow moiré effect has been developed by SIRA.[35] A section of a rotating radial grating is projected onto a small part of a specimen surface, and observed at an angle through the same grating. The resulting sinusoidal variation in light intensity is observed by a photocell. As the specimen moves away or towards the instrument, the phase of the photocell signal changes, and this can be measured with great precision by a standard electronic technique. A resolution of a few micrometres is possible with this instrument, and the use of phase measurements avoids the problems associated with amplitude measurements. The instrument was designed primarily for thickness measurements on fast moving plates and films.

Oster[36] reports two applications for studying refractive index changes. Two precisely aligned coarse gratings are separated by a distance (Fig. 28) and illuminated by a point source. The grating pitches are complementary, so that no light is transmitted. A transparent object, contained between two slides will deviate the light by amounts dependent on its refractive index, and a bright outline of the object appears at the second grating. A similar system can be used to study changes in the refractive index of water as a substance, such as sugar, is dissolved. A study of the index changes allows the dissolving mechanism to be determined. Theocaris[16] has devised a number of similar methods (which he refers to as 'remote gratings') to study the deviation of light rays by transparent objects which are loaded in their plane. The Poisson contraction will vary with the in-plane stress, producing changes in surface slope which act as prisms, deviating the light rays, and the moiré fringes are proportional to the partial stress derivatives.

In optics, the moiré effect has considerable applications. Imaging one grating onto another is a simple way of studying the distortion of a lens, as

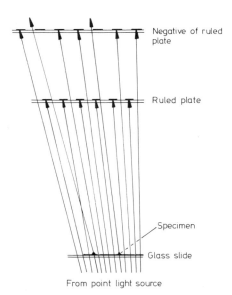

Negative of ruled plate

Ruled plate

Specimen

Glass slide

From point light source

Fig. 28. Lensless moiré microscope for studying refractive index changes on thin transparent objects (possibly biological specimens). (Courtesy of G. Oster.)

well as its resolving power, and gratings are used in instruments which test lenses on the basis of their optical transfer function (OTF). The moiré effect has also proved useful in Schlieren photographs and interferometry associated with wind tunnel and heat transfer problems. In Schlieren photography, the pressure gradients are visualised by photographic intensity changes. These changes can be quantified more readily by using the Schlieren optics to image one grating onto another (dispensing with the slot) and relating the gradients to the moiré fringe pattern. In Mach-Zender and associated interferometers, the refractive index changes are visualised by optical path differences. A problem with these interferometers is that large optical flats are used, and it is expensive to produce these with a flatness of 10 % of the wavelength used, which is the usual requirement for accurate interferometry. Larger tolerances produce systematic errors in the fringe patterns, but these can be removed by the moiré effect. The system is set up to produce a pattern of equi-spaced fringes, which are photographed. The system is then operated to produce the optical path length changes, and the new fringe pattern recorded. Superimposition of the two photographs produce moiré fringes which represent the optical path changes, the mirror defects being cancelled as they produce identical effects on the two interference patterns. The moiré fringes are identical to the interference fringes which would be observed if the mirrors were perfectly flat.

Visual observation of sharpened vernier fringes can be used for scale measurements between a long grating and an index piece,[37] to avoid the expense of photoelectric devices and their associated equipment. A mismatch is chosen between index and main grating to give fringes at convenient intervals, and this interval is divided into 10 parts to give a decimal readout (Fig. 29). Two moiré gratings are chosen to give the last

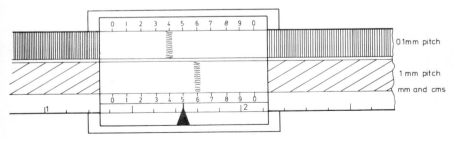

Fig. 29. Possible appearance of complete moiré vernier system.

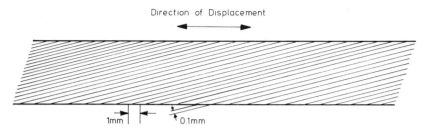

Fig. 30. Rotation of 10 lines mm^{-1} grating to obtain effective 1 line mm^{-1} grating.

two significant digits (0·01 and 0·1 mm) and the remaining digits determined from a conventional scale.

The coarse moiré grating in Fig. 29 has a pitch of only 1 line mm^{-1}, which is too easily visible to the eye, and obscures the fringe position. This is avoided by using a finer grating of, say, 10 lines mm^{-1}, rotated through about 90° to the measuring axis (Fig. 30), so that its effective pitch is 1 line mm^{-1}. The corresponding index grating is similarly adjusted, but with a different pitch, so that fringes are produced parallel to those of the finer gratings, and spaced apart by the same amount.

The system can be refined further by printing the index grating in the form of numbers (Fig. 31), so that the fringe position gives a digital readout. This is not entirely satisfactory, as the reading can be very confusing when the fringe is midway between two numbers.

This system was devised originally for use with two glass gratings plus an integral collimated illuminating system projecting the scale grating onto the index.[7] This allowed the two gratings to be separated by a small gap to prevent damage. A more convenient and cheaper system would involve a long metal grating, and glass index grating, with no integral illuminating system. This system could then compete economically with the conventional vernier system, but the gap between the two gratings would produce significant parallax errors in ambient lighting if the gratings were

Fig. 31. Appearance of grating with numbers superimposed.

not viewed perpendicularly to their surface. A simple anti-parallax device has been suggested to overcome this problem, based on the reflecting properties of highly polished metal gratings,[38] and this allows the direct replacement of conventional vernier devices by a moiré system at similar costs, but with greater sensitivity and convenience of use.

REFERENCES

1. Zandman, F., Holister, G. S. and Bracic, V. *J. Strain Anal.*, **1** (1965), 1.
2. Theocaris, P. S. *Moiré Fringes in Strain Analysis*, 1969, Pergamon Press, Oxford, chapter 3.
3. Fidler, R. and Nurse, P. *J. Strain Anal.*, **1** (1966), 60.
4. Vargady, L. O. *App. Optics*, **3** (1964), 631.
5. Burch, J. M. The metrological applications of diffraction gratings, in *Programs in Optics*, vol. II, ed. E. Wolf, 1963, North-Holland Publishing, Amsterdam.
6. Guild, J. *The Interference Systems of Crossed Diffraction Gratings*, 1956, Oxford University Press, London.
7. Guild, J. *Diffraction Gratings as Measuring Scales*, 1960, Oxford University Press, London.
8. Sciammarella, C. A. *Exp. Mech.*, **5** (1965), 154.
9. Purfit, G. L., Woodward, C. A. W. and Pettigrew, R. M. *NELEX Conference on Metrology*, Birniehill Institute, East Kilbride, 1974, paper no. 20.
10. Dew, G. D. and Sayce, L. A. *Proc. Soc.*, **A215** (1951), 536.
11. Burch, J. M. *Symp. on Optics in Metrology, Brussels*, 1958, Pergamon Press, Oxford, p. 361.
12. Williamson, D. T. N. *Machinery Lloyd (European Edn)*, **26** (1954), 51.
13. Davies, B. J., Robbins, B. C., Wallis, C. and Wilde, R. W. *Proc. Inst. Elect. Engrs.*, **107B** (1960), 626.
14. Shepherd, A. T. and Walker, G. S. Brit. Pat. No. 810 478 (1955).
15. Shepherd, A. T. UK Patent 1 525 049 (1976).
16. Theocaris, P. S. *Moiré Fringes in Strain Analysis*, 1969, Pergamon Press, Oxford, chapter 9.
17. Holister, G. S., Jones, W. E. M. and Luxmoore, A. R. *Strain*, **2**(4) (1966), 27.
18. Watson, J. and Holt, D. *Strain*, **2**(1) (1966), 17.
19. Dykes, B. C. *Proc. 4th Int. Conf. Exp. Stress Anal. Cambridge, 1970*, ed. M. L. Meyer, 1971, Inst. Mech. Eng., London, pp. 125–34.
20. Ligtenburg, F. K. *Proc. Soc. Exp. Stress Anal.*, **12**(2) (1954), 83.
21. Theocaris, P. S. *Moiré Fringes in Strain Analysis*, 1969, Pergamon Press, Oxford, chapter 5.
22. Post, D. and Baracat, W. H. *Ex. Mech.*, **21** (1981), 100.
23. Basehore, M. L. and Post, D. *Ex. Mech.*, **21** (1981), 321.
24. Luxmoore, A. R. In *Developments in Stress Analysis—1*, ed. G. S. Holister, 1979, Applied Science Publishers, Ltd, London, chapter 5.
25. Cook, R. W. E. *Optics Laser Tech.*, **3** (1971), 71.

26. Wadsworth, N. J., Marsland, M. J. N. and Billing, B. F. *Optics Laser Tech.*, **5** (1973), 119.
27. Graticules Ltd. Tonbridge, Kent.
28. Luxmoore, A. R. and Hermann, R. *Exp. Mech.*, **11** (1971), 375.
29. Shepherd, R. R. and Wensley, L. *Exp. Mech.*, **5**(6) (1965), 167.
30. Sciammarella, C. A. and Sturgeon, D. L. *Exp. Mech.*, **7** (1967), 468.
31. Sciammarella, C. A. and Rowlands, E. *Proc. 5th Int. Conf. Exp. Stress Anal.*, Udine, 1974, ed. G. Bartolozzi, p. 1.53.
32. Hitchings, D. J. and Luxmoore, A. R. *J. Strain Anal.*, **7** (1972), 151.
33. Davies, D. J., Evans, W. A. and Luxmoore, A. R. *Proc. Electro-Optics Int. Conf.*, *Brighton*, 1974, Kiver Communicators, Surbiton, Surrey, p. 51.
34. Pearce, I. K. In: *The Engineering Uses of Holography*, ed. E. R. Robertson, 1976, Cambridge University Press, p. 743.
35. Biddles, B. J. and Hobb, D. R. *Proc. Paper–Rubber–Plastics Automation Comp.*, Belgium, paper 13.1. (Reprints obtainable from SIRA Institute, Chislehurst, Kent.)
36. Oster, G. and Nishijima, Y. *Scient. Am.* (1963), 54.
37. Guild, J. *Scient. Am.* (1960), 198–206.
38. Luxmoore, A. R. *J. Phys. E. (Sci. Instr.)*, **1** (1968), 674.
39. *Optics and Lasers in Engineering* (Special Issue), **3**(1) (1982), 1–83.

Chapter 4

ALIGNMENT TECHNIQUES

P. W. Harrison

Hampton, Middlesex, UK

1. INTRODUCTION

Experience suggests that the seemingly explicit term 'alignment' is capable of varying interpretations depending upon the discipline in which one is working. For example, the surveyor might regard alignment of a road as setting out markers to define the path that it is intended to follow; this may be neither straight nor level. The engineer constructing very large electricity generating plant is much concerned with the relative alignment of the individual units of which it is comprised. Normally this is a straight line in plan, but in elevation it may more nearly resemble a catenary.

In this chapter alignment will be interpreted quite simply as the measurement of the departure of a series of points (or a continuous surface) from a straight line which may have any attitude; such measurements may be made in one plane containing this line or, commonly, in two such planes set at right angles. Measurements may be confined to assessing the fixed displacement of such points from the straight line or, in a constructional situation, may lead to their being moved on to that line if found misplaced. This is a quasi-static situation, but another possibility is that such displacements are varying with time, and may be the result of vibration, deflection or settlement of a structure. In this case it is their amplitude, and possibly frequency, which are to be measured. Other parameters which can be derived from the straight line, such as planeness and orthogonality, will also be discussed.

Two commonly met attitudes of the straight line are horizontal and vertical. Although over short ranges there is, in practical terms, no particular problem in the concept of a 'straight horizontal' line, strictly it is a contradiction in terms. The traditional levelling techniques used by surveyors and others employ gravity to define direction which effectively provides a reference which closely follows the approximately spherical form of the earth's surface and is not truly a straight line. There are engineering situations which clearly require one or the other condition. An interesting example is the form adopted for guide rails for the towing carriages which operate over ship model test tanks. These carriages are required to move at a very uniform speed; hence, although the rails must be straight in plan view, in elevation they must follow the earth's curvature because a 'horizontal straight' line goes uphill at an ever increasing rate at positions further away from the point at which the line is tangential to the earth's surface.

Human efforts to carry out alignment have been in evidence for many thousands of years, even if motives have been obscure at times. Two very early examples are the Nazca lines on an arid plateau in Peru, and the 'alignments' such as those at Menec in Brittany. The former largely comprise a series of straight-edged geometrical shapes and straight lines formed by removing loose surface rock; spread over a distance of 50 km, some of the lines are over 8 km long, crossing valleys and hills alike, and are said to have surprised surveyors by their precision. The Menec 'alignments' comprise 12 impressive parallel rows of granite blocks almost 1 km long. Their purpose is intriguing, and the techniques used in their construction certainly provide food for thought. The same can be said of the Egyptian pyramids built nearly five thousand years ago, followed three thousand years later by Roman construction of straight highways and towns having street patterns looking like the forerunners of the modern American city. Although many of the methods used by Roman surveyors were set down in a series of manuals, records relating to road-laying are unfortunately missing. However, at least one example exists of a *groma*, a device used for setting out the square road intersections in Roman towns.

The average town dweller has no need to look further than his garden fence or house wall to recognise that both are the result of a conscious effort to carry out an alignment, however imprecise. Indeed few, if any, branches of engineering can avoid the need to carry out alignment work, and examples can be found in the references drawn from the mechanical, civil and marine engineering industries in support of this contention. Refinement of constructional technique will often be linked to improved accuracy of measurement; this has been reflected in the design of new instrumentation and the adoption of new techniques to meet this challenge.

In recent years alignment has been particularly helped by the advent of the laser although it would be wrong to suppose that this has necessarily led to improved accuracy. Often it is the convenience in use that it affords the operator which has proved attractive.

2. INSTRUMENTS AND THEIR FUNCTIONS

When considering the evolution of alignment equipment it is clear that methods involving material references, such as taut string, wire or nylon, while having a useful part to play, particularly for alignments in the horizontal plane or in the vertical direction (for example, the plumb bob), nevertheless are limited on range in normal usage, and suffer to some degree from external influences such as draughts and gravitational forces.

The adoption of 'weightless' optical references at once seems attractive both in terms of straightness of reference and of the range which they promise to provide. However, this promise will only be fulfilled if the operator is well-informed on practical aspects of the functioning and handling of his instruments, and has an awareness of possible pitfalls in applying the technique he proposes to use. Although it is not proposed to consider the design of each instrument in detail, an understanding of its basic functioning is necessary to appreciate the sources of error which can be present, and the way in which they can best be minimised by adoption of suitable procedures and techniques. This chapter now sets out to attempt to provide helpful information on these two aspects.

The first group of instruments to be examined embraces those in which a telescope defines the optical reference.

Galileo is sometimes wrongly credited with inventing the telescope in about 1612; others have questioned whether the knowledge of the ancients could have been acquired without the assistance of some such device. The writings of Bacon certainly indicate his awareness of the theory on which the telescope operates. However, the practical discovery of the telescope was made in Holland in about 1608; it was Galileo who did much following this to improve telescope performance, finally achieving a magnifying power of 33. In 1611 Kepler pointed to the advantages of a telescope constructed of two convex lenses, but it was Gascoigne who recognised the practical possibilities of an instrument in which a positional reference object placed at the common focus of the two lenses can be viewed simultaneously with a distant object. The ensuing story of the telescope is fascinating[1] but lengthy; Gascoigne's micrometer, which was incorporated in astronomical telescopes so that the diameters of celestial objects could be deduced, was a significant step in the development of the telescope as a measuring device.

In its simplest form (Fig. 1) the telescope comprises two positive lenses separated, approximately, by the sum of their focal lengths. A real image of a remote object will be formed close to the focal plane; because of its bias towards the eye lens an enlarged but inverted virtual image will be seen by the eye. If a cross-wire is placed in the focal plane with the intersection on the common optical axis of the two lenses, then the eye will see it superimposed on the image. All points which appear to be at the

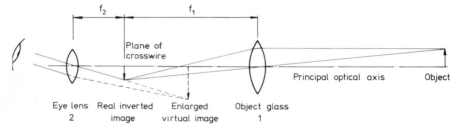

Fig. 1. Simple form of terrestrial telescope.

intersection would then be known to lie on the principal optical axis of the instrument.

The modern terrestrial telescope is necessarily a much more highly developed instrument in which the image is usually erect and the lenses are corrected for chromatic and other aberrations. Axial positioning of the eye lens usually serves only to bring the graticule into focus, this being followed by a second focusing operation carried out by moveable lenses placed between object and eye lens which enables objects at varying distances away to be brought into sharp focus in the plane of the graticule, and hence in focus with the observer's eye. Furthermore, when an object viewed through the telescope is seen to be displaced laterally from the cross-wire intersection, this displacement may often be measured with the use of an optical micrometer. This device may be incorporated into the instrument itself or provided as an attachment. Basically it comprises a flat parallel glass plate (Fig. 2) which, when rotated, causes light rays entering it to continue, upon leaving, in the same direction but displaced laterally. In

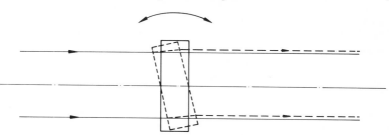

Fig. 2. Principle of optical micrometer.

this way an object appearing off-axis may be made to appear on axis by rotation of the micrometer; the rotation necessary is a measure of the offset. Depending upon the type of instrument, the micrometer may measure in one direction or also, by tilting about an axis at right angles to the first, in two directions. The diagram in Fig. 3 shows the ray path through a modern alignment telescope incorporating some of the features referred to earlier.

Today there are numerous instruments commonly used for some aspect of alignment work in which the telescope is an essential feature. Firstly, the optical level consists of a focusable telescope, with a magnifying power typically between 20 and 30 (occasionally more), which is so mounted that its optical axis will be brought closely horizontal when the bubble of a sensitive spirit level, to which it is rigidly attached, is centred. The telescope can be rotated on its base about a nominally vertical axis and can therefore be made, after relevelling, to generate any number of optical reference lines all lying in the same horizontal plane. The relevelling required for each pointing is eliminated in automatic levels which are so designed that, once the instrument base is approximately levelled by reference to a small circular bubble, the optical axis is automatically brought closely horizontal

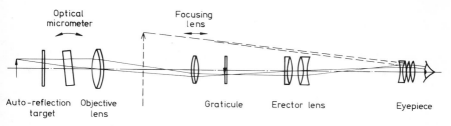

Fig. 3. Principal optical components in Taylor Hobson microalignment telescope.

by the action of gravity upon compensating optical components within the instrument. This feature facilitates setting up, and small tilts of the instrument support do not give rise to tilt of the optical axis. Although in a surveying situation it may sometimes be adequate to sight on to a remote graduated staff, the potential for high performance found in the better instruments can only be realised in engineering situations if the scale viewed is itself mounted on a vertical micrometer slide. A line on that scale can then be set fiducially on the instrument graticule, but requires two observers to accomplish this. Alternatively, the instrument is fitted with an optical micrometer of such range that it can always be sighted on to one or other line of the scale in a fiducial manner. Fiducial setting of this sort is something at which the human eye is very good, certainly better than its estimation of a fraction of a graduation interval.

Secondly, the optical plummet performs the function in the vertical direction that which the level carries out in the horizontal. Once again manual or automatic means are used to bring the axis of a telescope to the horizontal, but it now views through a 90° constant deviation prism (described more fully later) which may either direct the sight line upwards or downwards. Moreover, in one example of a modern instrument, the sight line is brought to the vertical automatically in two orthogonal planes using corrective techniques similar to those used in the level.

Thirdly, the theodolite again employs a focusable telescope (many have a magnifying power approaching 30) which is supported in a precision trunnion enabling the telescope to be elevated and depressed and, indeed, transited (turned completely over); the entire trunnion and telescope assembly (the alidade) can be rotated about an axis normally set vertical with the aid of a sensitive bubble incorporated in the alidade. Scales incorporated in the instrument enable elevational and azimuthal (rotation about a vertical axis) angular movements to be measured. Hence the optical axis of the telescope can be pointed in nearly any direction, including vertically upwards in instruments fitted with suitably angled eyepieces, its pointing being defined by reference to the scales. The elevational scale is customarily related to gravity, and in some recent instruments this is done by automatic means. Use of a theodolite in attitudes other than that described generally requires special provision in the load-carrying bearings if performance is not to be unacceptably degraded. When two theodolites at a known separation are used to sight on to a series of targets it becomes possible to carry out three-dimensional measurements.

These three types of telescope-based instrument, formerly regarded as the tools of the surveyor, are becoming increasingly used in the mechanical engineering industry. As a result they may be handled by staff who have not had the training of the surveyor. In contrast, the precision alignment telescope (the best known example being the Taylor Hobson instrument, first made in 1940) has been based in that industry for many years and continues to serve it well. The experience of Taylor Hobson has subsequently led to improvements in design and breadth of application made possible by the development of a wide range of ancillary equipment to the extent that their instrument can now perform additionally some of the functions of the other types of instrument referred to here.[2]

A technique which uses these instruments to minimise measurement errors is discussed later in the chapter.

3. THE LASER AND ITS IMPACT

It is proper and natural that many engineers should be concerned to know what contribution the laser can make to the processes of alignment; for some it still seems a somewhat exotic tool to employ, far removed from taut string. It is not true, unfortunately, that laser light has the ability to travel in straighter lines than light of the same wavelength from other sources. Nevertheless laser-based systems are rapidly becoming established in the measurement field because the laser does have other distinctive characteristics which enable tasks to be undertaken that could not have been contemplated before its inception.

If certain fundamental aspects of laser behaviour are understood it is reasonably straightforward and inexpensive to design and use laser-based systems for many measurements in a very effective manner. There are three principal ways in which a laser may be used—firstly, its beam may serve to provide a visible reference, replacing the optical reference axis of the more traditional instruments, and as such defining both a position and direction in space, and secondly, as in the technique extensively used by the National Physical Laboratory (NPL),[3,4,5] the laser acts only as a bright light source defining one point on an optical axis. Thirdly, the laser is used as a bright light source in conjunction with conventional optical instruments for the projection of its beam along their optical axes. (A fourth and technically more complex approach[6] is available which, unlike the ways just described, uses the quality of coherence of laser light in a length-measuring inter-ferometer so adapted that it can measure lateral displacement from a straight line.)

The laser commonly used in these applications has a helium–neon filled plasma tube which emits continuously (some lasers are pulsed) a highly monochromatic red beam of wavelength 0·6328 μm. Typically, on leaving the laser, its diameter will be between $\frac{1}{2}$ and 1 mm. Reference in the literature to the $1/e^2$ diameter of the beam requires some explanation. The laser beam is not uniform in intensity (power per unit area) over its cross-section; it has a Gaussian distribution (Fig. 4), similar in form to a random error frequency distribution curve, and the specified diameter is the beam width at which the intensity has fallen to $1/e^2$ or 0·135 of its peak value.

By the natural laws of light propagation the beam, although it may well be collimated on leaving the laser, will soon begin to diverge at a rate which is related to its initial diameter. This could mean that, used as a reference over a considerable distance, say 100 m, the beam could have expanded to a diameter of at least 100 mm. There are two drawbacks to this situation; firstly, the laser power is then spread over an area many thousands of times greater than at source, and it may be difficult to observe the beam in bright ambient conditions, and secondly its size would make the assessment of the position of its centre much less precise when, for example, it was allowed to fall on some form of target. Fortunately, this situation can often be overcome by using a beam expander in front of the laser. The rate of divergence of the beam is then reduced in the same ratio as beam diameter has been increased; for example, a beam having a natural divergence of 1 milliradian (one part diameter in a range of a thousand) passed through a 10 × beam expander can be made to diverge at only 1/10 milliradian.

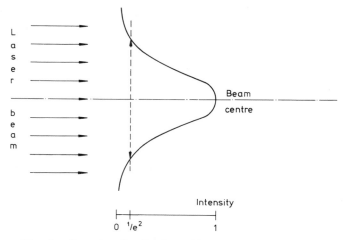

Fig. 4. Gaussian distribution of intensity in laser beam.

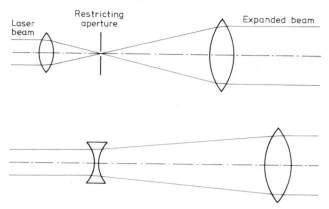

Fig. 5. Two forms of simple beam expander.

Beam expander design is reminiscent of the telescope, consisting in essence of two positive lenses or a negative and a positive lens (Fig. 5). Although the second design is more commonly found, the first is sometimes used when a particularly 'clean' beam of much magnified diameter is required because it can be spatially filtered, a process consisting of placing a small circular aperture (possibly no more than $30\,\mu m$ diameter) in the common focal plane of the lenses. Its function is to remove from the light reaching the objective lens all rays which either arrive at the entry lens in a direction slightly different from the main stream or, because they suffer internal reflection within the entry lens or are diffracted by dust on the lens, behave as though in the former category. Such rays, if allowed to pass through, could give rise to interference fringes in the beam or to other imperfections in beam appearance; because they arrive at the filter off-axis they fail to pass through the aperture. Powerful beam expanders may use a microscope objective in place of a simple entry lens, and some expanders have provision for the operator to focus the beam so that it is not necessarily collimated on leaving the expander. The latter facility may be used when, for example, in seeking to achieve a beam having the smallest possible diameter over the selected range, the diameters at source and at the far end are to be made the same. The beam then takes the form shown in Fig. 6 in which the diameter at the neck is $1/\sqrt{2}$ that at the ends. A simple formula relates initial diameter, d, to the range, L, namely

$$d^2 = 0\cdot 80L$$

where d is in millimetres and L is in metres.

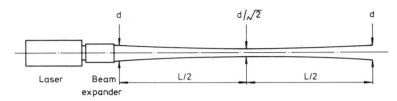

Fig. 6. Beam form for minimising diameter variations.

The precision of the reference so generated by a laser relies upon the constancy of laser pointing during the alignment process. Any disturbance of the instrument, whether caused by handling, support movement or refocusing of the beam expander, will result in an angular shift of the optical axis and the likelihood of an unacceptably large positional error of the axis at large distances from the instrument.

To some extent this possibility may be countered when using the laser for simple alignment work by adopting an approach similar to that used with the alignment telescope, namely to locate a reference target at the far end of the alignment path such that the straight line from laser to target origin represents the required reference. By directing the beam at this origin and then interspersing alignment of intermediate points with reference back to this target a check can be maintained on laser pointing; however, the assumption is made that no pointing changes occur between checks.

Specific instances of the use of the laser in the first of the three ways listed are largely found in civil engineering applications. There are numerous examples in which the laser suitably housed for field use and often with special support arrangements can be set to define a horizontal reference or a reference at a predetermined inclination. Use is generally made of a sensitive bubble to define the horizontal, with inclination being generated by mechanical means. One commercial version holds its inclination automatically once set.

For the generation of vertical references there are instruments in which the laser in some mounting is brought vertical manually by reference to two sensitive levels orientated to indicate tilt in two orthogonal directions. In other versions displacement of the level bubbles from a central position in their vials causes an imbalance in an electrical circuit which then drives the laser and its level-carrying base to its correct attitude. A simple but effective design of automatic plumb (Fig. 7) developed at NPL[7] employs a laser and its small power pack suspended from a trunnion mount such that, although the supporting body tilts, the laser always hangs vertically, rather

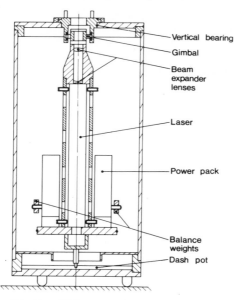

Fig. 7. NPL upwards-pointing plumb.

like a plumb bob and line, and directs its beam, in the version illustrated, upwards. The 'pendulum' is damped by a probe dipping into a viscous liquid to make it insensitive to vibration and a special feature is the ability of the plumb to be self-calibrating by rotation of the trunnion support through 180°. Constancy of beam position when viewed some distance away is an indication of correctness of pointing.

Another way in which the axis-defining laser may be used is to generate a horizontal reference plane. To appreciate the manner in which it can do this reference must be made to the constant deviation prism, a device which the engineer building up his own system may often find helpful in other contexts. Figure 8 shows one example in which the angle between entry and exit rays has been made nominally to be 90° although other angles are possible. As its name implies the prism may be rotated in the plane of the paper without disturbing this angular relationship. Unwanted tilt about the X–X axis can, if not adequately controlled, give rise to very small variations in the deviation angle; for a 90° deviation angle the variation equals the square root of the tilt. Several manufacturers have used its qualities by mounting it above a laser beaming vertically upwards, the emerging beam then being horizontal. By arranging for the prism to rotate

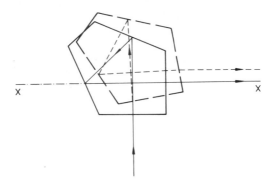

Fig. 8. 90° constant deviation prism.

about a nominally vertical axis, either motor-driven or by hand, the beam sweeps out a horizontal plane. In a situation in which accuracy of performance requires a beam pointing constant to a few seconds of arc, the use of a prism of this type makes the behaviour of its rotation bearing much less critical. In NPL designed equipment used for monitoring deflection of the crest of an arch dam use has been made of another form of constant deviation prism to generate a 'bent' optical axis.

The approach adopted by NPL which constitutes the second way in which the laser can be used is best understood by reference to Fig. 9, study of which explains the choice of name, three point alignment. Here the laser is introduced into an alignment system in which its sole function is to act as a point source of intense light, and it now defines only a point through which the optical axis passes. The method then adopted to completely define the axis is to allow the light from the laser to illuminate a remotely sited image-forming device (such as a glass lens, Fresnel zone plate, or coarse grating), the centre of which identifies a second point on the axis.

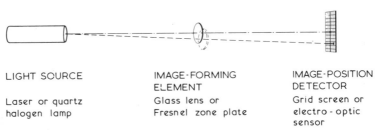

LIGHT SOURCE	IMAGE-FORMING ELEMENT	IMAGE-POSITION DETECTOR
Laser or quartz halogen lamp	Glass lens or Fresnel zone plate	Grid screen or electro-optic sensor

Fig. 9. Principle of three point alignment system.

The important feature of such an arrangement is that the axis-defining units do not have to be set truly perpendicular to the axis and that small angular movements of these units can be tolerated without introducing significant errors into the axis position. The axis can be detected further along the optical path by receiving the axis-centred image on a screen. The system therefore comprises three units, and because many problems can be interpreted as a need to measure, at any one time, the alignment of three widely separated points, this can be done by relating the units to these points in a defined manner, usually by mechanical means. If it is arranged that when these points are in line the image is at the screen datum, then displacement of the image from this datum becomes a measure of the misalignment of the points, both in the horizontal and the vertical plane. Both lenses and Fresnel zone plates, but not gratings, have the quality that the in-focus image has an intense centre which makes it particularly suitable for viewing when the ambient lighting is bright, and is highly desirable when image position is to be sensed electrooptically. Zone plates (Fig. 10(a)) do not have the limitation in focal length imposed on glass lenses by manufacturing considerations. The coarse grating (Fig. 10(b)), favoured by some in The Netherlands where the basic three point concept has been exploited for many years,[9] forms an image continuously over a considerable axial distance with changes in image proportion rather than form; by contrast the image formed by a zone plate at axial positions away from the image plane, while still having a readily identifiable centre changes its appearance considerably with axial position. In all cases, of course, the

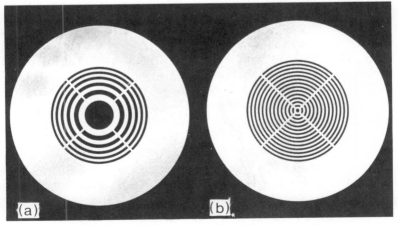

Fig. 10. (a) Fresnel zone plate; (b) coarse grating.

image is circular, and, when using zone plates and gratings, has light and dark rings surrounding the image centre. However, the power distribution in the grating image is far more uniform and the centre consequently much less bright. The laser beam used in this system should not be expanded in the manner already described, but may sometimes need to be diverged using a weak negative lens in order to overfill the imaging element with its light.

A third way of using the laser is in association with instruments through which the observer normally views the target. This procedure has been applied by at least two manufacturers to instruments which include the level, plumb and theodolite. Their approach has been to mount a helium–neon laser typically of 5 mW power on the leg of the tripod supporting the instrument, light being fed into the instrument by fibre optics through a special eyepiece

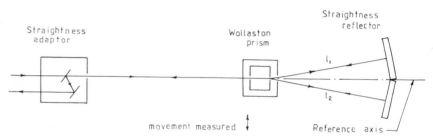

Fig. 11. Adaptation of interferometer to measure straightness.

which replaces the conventional one. The telescope of the instrument then behaves rather like the beam expanders already described, and by focusing the light emerging from the fibre to a minute point centred on the reticule, the beam is made to coincide with the axis defined by the instrument in normal usage. Although a considerable proportion of the laser power is lost in the fibre optics and elsewhere, safeguards in the form of filters are provided to permit simultaneous observation through the instrument by eye.

A fourth approach uses the laser in a length-measuring interferometer which compares two optical path lengths, l_1 and l_2 (Fig. 11), between the beam-dividing Wollaston prism and two precisely flat mirrors set at a fixed angle. Lateral movement of the prism with respect to an axis defined in one plane by the bisector of this angle produces a proportionate change in $l_1 \sim l_2$ which is displayed.

3.1. Detection of the Position of a Laser Beam (or Laser-Formed Image)

When using traditional survey instruments the observer will have a considerable choice of commercially prepared targets or scales suitable for viewing over the larger ranges. The choice for short range work may be more limited, sometimes requiring that special targets be tailor made to the particular need.

With laser-based instrumentation, for example, the need is to identify the position of the beam (or image centre) with respect to some datum. Crossed lines drawn on card may often suffice for relatively coarse centring, but for more precise work or when measuring the offset from a datum a simple translucent grid screen has been found most effective. This can comprise a proprietary grid-marked plastic sheet glued to a glass or Perspex backing, and can be used in either of two ways. The eye can assess displacement from an origin by reference to the grid on the screen; when used this way, for optimum precision, an 0·1 inch square grid has been found best, and is less likely to lead to numerical mistakes than, say, a 2 mm square. However the eye is better at judging symmetry and the greater accuracy will be obtained if the grid screen is moved by micrometer until the screen origin or a selected intersection is centred on the beam. Some observers have expressed a preference for a screen comprising concentric circles for this approach, but NPL has no evidence of a resulting improvement in accuracy.

The advantage of a translucent screen is that when beam power is low and ambient lighting relatively high the spot of light will be seen more clearly when viewed through the screen, especially if the latter is fitted with shielding.

There are three types of situation when it is not feasible to use the combination of human eye and screen in any of these ways. They are when making alignment measurements in a dynamic situation which causes the beam to move too quickly for the eye to follow and record, when detection is to be made in a position inaccessible to or hazardous for the observer or, thirdly, when observations are to be made over a long period, perhaps days. These circumstances will often be appropriate for electrooptic detection. There are several different types of position-sensing detector, and certain basic factors must be considered when adopting one or other system. The observer is only interested in studying the behaviour of laser-radiated light unadulterated by ambient lighting, whether daylight or artificial. The latter can have the effect of altering the power distribution across the detector causing a false assessment of beam position, or it can so raise the level of power falling on the detector that it reaches a 'saturated' condition in which it will no longer respond correctly to beam movement. It may suffice to place a filter in front of the detector which, while allowing the laser light to pass, will eliminate all ambient light other than that in a wavelength band close to that of the laser radiation. Experience at NPL indicated that interference filters were insufficiently uniform over their working surface for this purpose, and two dyed-glass filters were adopted, their combined characteristics providing a very narrow transmission band. In more adverse conditions it may be necessary to modulate the laser beam, a process in which the beam at source is chopped at high frequency (in the NPL system about 1 kHz) usually by mechanical means. The laser power arriving at the detector can then be distinguished from the ambient lighting, which will generally be continuous or modulated at twice the relatively low frequency of the mains supply, by designing the electronic circuitry of the detector to respond to frequencies close to the chopping frequency and to ignore all other signals fed to it.

The electrooptic detector, like the grid screen, may be used to measure displacement from an origin, or it may be moved until centred on the beam, that movement being measured by micrometer or other means. The quadrant detector, which measures displacement in two axes, comprises four cells arranged as in Fig. 12. Typically, opposed quadrants are electrically linked, and as the beam falls on them the associated electronic circuitry measures the differential photoelectric current arising when the power level on one is higher than the other, that is, the beam is displaced

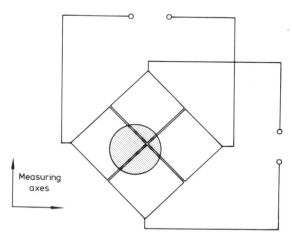

Fig. 12. Quadrant cell configuration.

from the centre. If the beam passes into one quadrant of a pair only, the detector ceases to record movement in that direction, and is in any case increasingly nonlinear in its response as this situation is approached. An improved way of operating a quadrant cell has been described by Bennett and Gates.[10]

Another form of detector uses a position-sensing Schottky barrier PIN photodiode. The two-axis version uses a single chip having four current electrodes spaced equally around the periphery. When light falls on the cell current is generated which apportions itself between these electrodes according to the respective path resistances, and hence distances, between the light spot and the electrodes. The function of the associated electronic circuitry is therefore to monitor and, after suitable processing, display or record the current difference between the two pairs of opposed points. In a single axis version the sensing element comprises a narrow strip which may be up to 100 mm long. The former is, once again, inherently somewhat nonlinear when used over its maximum sensing range, but the latter is much better in this respect. Some lack of linear response can occur due to local irregularities in cell behaviour, but these will be minimised if advantage is taken of the beneficial averaging effect of using a light spot which is no smaller than, say, 2–3 mm diameter.

In the detectors described any change in power level in the beam will give rise to varying differentials for a given displacement, so changing detector sensitivity. It has therefore become common practice to adopt the approach found in the NPL circuitry[11] in which the total signal from each pair of pick-off points is divided into the respective difference signals. Change in power level leaves this ratio unaffected, and the system is said to be fitted with AGC (automatic gain control).

Such effects and problems of nonlinearity become unimportant if the detector is always moved to a null (electrically balanced) position with respect to the beam, other means being used to measure the displacement of the detector. In certain applications the use of a small hand-operated micrometer coordinate slide upon which to mount the detector will commend itself, especially when a range of movement in excess of that of the cell is required. However, from inertial and other considerations, this may be impracticable in a dynamic situation. A better response time would be achieved by adopting servocontrolled optical means responding to the out-of-balance signals which displace the beam to the null position on a stationary detector.

Two further practical points should be made. The narrowness of the sensing area of the single axis Schottky diode means that a beam having a small diameter could, if subject to a lateral drift, wander off the diode. In the NPL detector this possibility is minimised by placing a 30 mm wide cylindrical lens of about 55 mm focal length at this distance in front of the cell, with its axis parallel to the detector measuring axis. The lens both directs the beam on to the diode and condenses it in the lateral direction, but has no effect along the axis of measurement. Secondly, in order to increase the effective range of the two-axis diode a large aperture lens (conveniently a commercially-available acrylic Fresnel lens) can be placed in front of the diode so that its focal plane lies somewhat behind the diode. This reduces beam movement in the diode plane and if, for example, diode to focal plane distance were one quarter of the focal length, then image movements four times greater than the normal range could be monitored. However this ability would be accompanied by some degradation in sensitivity, and the reading does become sensitive to tilt of the lens–diode combination. In a situation in which a high power beam (or image) is being reduced to a smaller area at the diode, extra care must be taken to avoid intensity (power per unit area) levels which could damage the diode. The manufacturer's guidance on acceptable values should be sought.

Various ways are adopted to display the information on beam position. When static or slowly changing values are being observed, a meter readout may suffice. However, for faster changing events and long-term monitoring the use of a chart or tape recorder will be necessary.

The atmosphere through which the light is passing can cause the reading to flutter at a frequency which will often be considerably higher than that of the movement being monitored. In order to counter this the NPL circuit incorporates a filter which provides a choice of electrical time constants that can be introduced to reduce the unwanted effects. Other significant effects of the atmosphere not removable in this way are dealt with elsewhere.

4. APPLICATIONS AND ANCILLARY EQUIPMENT

So far this chapter has indicated in broad outline the types of instrumentation available to the engineer faced with a problem of measurement. Choice of instrument and technique and the design of ancillary equipment will now be further illustrated by reference to a few specific tasks.

4.1. Theodolite

The example cited has been selected because use is made of measuring abilities peculiar to the theodolite.

In connection with the testing of ship models it was required to measure the relative position in space of four small lights situated above a 30 metre square water-filled tank. The survey therefore had to be made from the tank edge, and used two tripod-mounted theodolites. The smooth concrete floor necessitated using 'spiders' in which to locate the tripod feet. Theodolite separation was determined by using the in-built optical plummets to view graduation lines on a carefully tensioned invar tape laid on the horizontal ground and positioned directly beneath the theodolites. The very low coefficient of expansion of invar (usually less than 0.5 ppm $°C^{-1}$) makes such a tape relatively insensitive to temperature changes.

The process of triangulation comprised two principal steps, firstly the measurement of separation of the lights when their positions are projected on to a horizontal plane, and secondly assessment of their heights above a selected datum. For light A (Fig. 13) the theodolites were used to measure angles α_A and γ_A, and this led to the calculation of its coordinates in plan with respect to T_1 as origin and T_1–T_2 as direction. These angles were obtained by sighting each theodolite at the other using the conical backsight on the telescope as a target, and noting the azimuth readings.

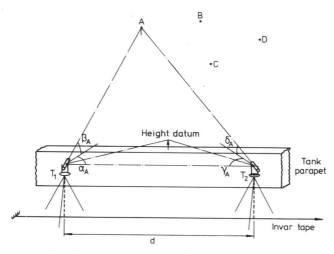

Fig. 13. 3D survey with theodolites and tape.

Both telescopes were then swung round to view lamp A, the change in azimuth reading corresponding to α_A and γ_A. At the same time the elevational angles β_A and δ_A were noted after careful setting of the bubble associated with the elevational scale. The same observational data were obtained for the other lamps, and elevational sightings were additionally made on a special height datum set up on the tank parapet. The combination of α and γ values with theodolite separation, d, provided values of lamp positions with respect to the theodolites and hence their relative separations in the horizontal plane, and the association of elevational values with the respective horizontal separations led to the determination of height differences between each lamp and the parapet datum. It should be noted that each theodolite generated its own set of height values, the agreement between which was a pointer to the accuracy of the survey. This agreement was typically 0·2 mm in an array having lamp separations of several metres.

Because sighting distances varied telescope refocusing was inevitable, and shift of the optical axis became a possibility. The very standard practice was therefore adopted of repeating all observations with both telescopes transited; the averaged result then largely eliminated errors arising from this and other significant causes.[12]

4.2. Automatic Level

When setting out heavy engineering plant a level in conjunction, perhaps, with an instrument such as the alignment telescope or a technique such as the three point system, may be applied directly to references on the machinery; alternatively datums grouted into the floors may serve as intermediate references. Automatic levels of appropriate grade have been found both accurate and speedy for such work. Normally, for the convenience of the observer, the instrument is mounted near eye-level; it is then necessary to relate the scale to be viewed to the ground reference below. In early trials at NPL such datums were provided with a 90° countersink, a ball-footed staff being used to carry the scale; the scale graduations were spaced at 10 mm intervals matching the range of the optical micrometer on the level. The ball was formed on the end of studding screwed and locked in a threaded insert in the end of a brass tube carrying at its upper end a cross-levelling bubble and a recessed ring into which a disc carrying the scale was clamped. The choice of a standard and precise recess diameter of 120 mm provided versatility, the staff being enabled for other applications to support alternative disc-mounted items such as imaging elements and target screens. It is a good principle in such survey work to site the level so that variation in length of sight lines and the consequent need to refocus is a minimum. Errors arising from the sight line generated not being truly horizontal or becoming curved due to atmospheric refraction are minimised. However the separation of the positions to be surveyed could be such that to observe all from the one position would result in excessively long or obstructed sight lines. It is then necessary to resite the level at least once so that, if possible, not less than two of the positions surveyed from the first location form part of the second stage of observations, this establishing more certainly the relationship between the two sets of observations. When the points to be surveyed lie in a long straight line an effective procedure is to mount the level mid-way between the first two, sight back to the first then forward to the second, repeat the observation and return to the first. This provides a balanced set of four observations. NPL procedure[3] was to use two staves in which the heights of the scales above the ball-feet had been made closely equal by appropriate rotation of the studding carrying the feet. The level was then moved up to view the second and third points, the staves being advanced in leap-frogging fashion, proceeding in this way to the end of the line and returning in a similar manner. The summation of the height differences for the complete cycle should equal zero; the uncertainty of the survey was reflected in the size of the departure from zero. In this particular exercise ten points spread over 750 m were surveyed in this metrologically rigorous manner, and repeated cycles of observations indicated that the assessed positions of stations 2–9 with respect to stations 1 and 10 had, in the mean, a spread of 0·3 mm. Even better performance could be expected over a shorter range.

4.3. Alignment Telescope and Three Point Alignment System

The installation of a new Van de Graaff accelerator at NPL required that a number of major items of equipment were set out along two horizontal straight lines extending in both directions from a datum defined for position and, in the shorter limb, direction by the opposing pole pieces of the analysing magnet. In order to achieve this the Taylor Hobson alignment telescope was used on the shorter limb in conjunction with a reflecting target, the latter being mounted on the magnet at the correct height and lateral position, and orientated in plan to be square to the desired direction; in elevation it was nominally vertical. At a position along the line so defined but lying beyond the accelerator component most remote from the magnet the telescope was set up on a proprietary support, providing pointing adjustments, which was itself mounted on a robust double cross-slide. By focusing the telescope to infinity and using it as an autocollimator[2] its axis was brought square to the reflecting target in plan and horizontal by use of a stride level mounted on the telescope. The telescope was then focused upon the target and moved by cross-slide until its crosswires were centred on it. Accuracies of 0·1 mm on position and a few arc seconds on attitude were sought and readily achieved with a telescope-to-magnet separation of 4 m.

This procedure made the telescope axis coincident with the required line, and it then became a relatively simple task to bring the various units of the accelerator which formed that limb on to the telescope axis by viewing targets suitably mounted in them. Whenever possible reference was made back to the reflecting target to confirm that the telescope position and pointing had remained unchanged.

The other limb of the assembly was nearly 20 m long, and the required reference line was defined both for height and lateral position at one end by a burn mark on a quartz plate at the output port of the analysing magnet, and laterally at the other by a small countersink in a post set at ground level. An automatic level placed equidistant from these two points was used to sight on to a target laid over the burn mark, and to transfer that height to a target in a ball-footed staff of the type already described which was located in the countersink. The staff was adjusted to bring its target to the same height as the burn mark, and then carefully cross-levelled to position the target vertically above the foot. Having thus identified the end points the three point technique was used to align the intermediate units of the system in the following manner. The target on the staff was replaced by a

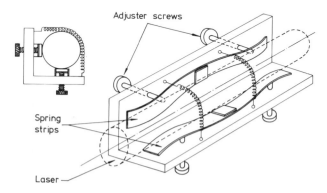

Fig. 14.   Simple NPL-designed laser mount.

coarse grating, and a tripod-mounted low-powered helium–neon laser was so positioned that the image formed after the beam had passed through the grating was centred on the burn mark: the optical axis thus coincided with the prescribed line in this limb. By using appropriate adapting pieces a centralised target was located in the various units in turn, unit position being adjusted until the image became central on the target.

The use of the laser in this way requires that it is so mounted that it can be both finely pointed and positioned. A simple inexpensive but very effective mounting suitable for cylindrical lasers has been made and widely used by NPL on many occasions; Fig. 14 shows the main features of its design. The laser is held by springs against two shaped strips of spring steel, each secured centrally but operated upon by a pair of lockable screws. Rotation together translates the laser, differential rotation changes pointing, and the design ensures very little 'cross-talk' between horizontal and vertical movements.

This task was a particularly good example of the complementary nature of the methods used.

5.   ATMOSPHERIC EFFECTS

Certain conditions of the atmosphere can affect both straightness and steadiness of optical paths. Perturbations caused cover a wide band of frequencies and give rise to a situation in which the higher frequencies become readily apparent when, say, a light beam is received on a screen; then either the eye or electrooptic detectors can assess its mean position. However other frequencies may be so low as not to be apparent during the course of the observational work, and could go undetected. The cause of such perturbations lies in the presence of innumerable pockets of air lying along the optical path which give rise to fluctuations of refractive index, itself dependent upon temperature, pressure and humidity; because light bends towards regions of higher refractive index any transverse gradient will act like a weak prism in the light path, and the summation of such effects may cause the beam to move first one way and then the other. However, in certain conditions, such as when a path passes above ground warmed by the sun, the overriding factor can be a vertical temperature gradient that may only change slowly with time. Its presence is best detected using a vertical array of temperature sensing elements.[3]

The observer may be helped in assessing the significance of long-term bending in relation to the required instrumental performance if it is remembered that a transverse temperature gradient of $1\,°C\,m^{-1}$ present at all points along the optical path will cause a beam to deviate from a straight line by, approximately, 5 mm at 100 m and in proportion to the range squared for other distances. Hence, at 10 m the figure is 0·05 mm. By comparison the ever present vertical gradient of pressure in the atmosphere would cause the beam to bend by only 0·008 mm in 10 m. The use of an evacuated light path to eliminate the error is seldom practicable in an engineering environment, and a static air-filled pipe, while damping the readily observed oscillations, will often aggravate long-term errors due to increased stratification of the air in the pipe. Circumstances may therefore require the observer to be selective as regards time and, possibly, sighting paths in order to reduce such effects to an acceptable level. NPL experience suggests that a visibly oscillating image is probably also subject to long-term displacement, and that the extreme condition of a very fragmented image will commonly mean that conditions are changing too quickly for reliable measurements to be made; indeed the observer is unlikely to be able to identify a centre to the image. Other approaches to the problem have been the subject of some investigation at NPL[13] and include the use of slowly rotating pipes to eliminate transverse gradients. When the three point system is applied to a large machine tool, the imaging element being mounted on a moving carriage, this approach is not viable. A scheme found effective was to use a stationary tube having a slot along its length through which an arm could be passed to support this element within the tube section. The air in the tube was stirred by air entering it through orifices in a pipe laid inside the tube. Reduction of gradients to a quarter or less of those in the ambient air were achieved.

When working in an outdoor environment a heavily overcast day with some wind, and even light rain, will often provide the most steady and refraction-free sighting conditions.

6.   LASERS AND SAFETY

This subject has received considerable attention, and the safe levels of power and acceptable modes of operating are becoming more widely agreed. These are shortly to appear in a significantly revised edition of British Standard 4803, 'Guide on protection of personnel against hazards from laser radiation'. While the user should refer to the full and current text as the proper source of authority, certain salient features of the revised edition may be noted. For example, present thinking is that, when using a helium–neon laser not exceeding 1 mW in power, no special eye precautions are needed, sufficient protection being afforded by aversion responses, including the blink reflex, and that *accidental* viewing is not hazardous even using optical aids. That is not to say that it would not be disconcerting for personnel to find themselves accidentally viewing such a laser straight in the eye, especially at close quarters or at any position where beam diameter remains small, and common sense suggests that this situation is best avoided if possible. Above this power and up to 5 mW for a helium–neon laser, some reliance is placed upon the same aversion responses to limit the time that the eye would allow itself to be exposed to the radiation, but additionally and importantly effective steps must be taken to make it impossible for accidental exposure of the eye to the beam until it has been so expanded or diverged that the mean irradiance (intensity) over the central 7 mm diameter does not exceed 25 W m^{-2}. The beams of a 2 mW and 5 mW helium–neon laser would therefore have to be increased to 12 mm and 21 mm diameter, respectively, to reach this threshold. It could still be hazardous if, for example, some optical viewing device was placed in the beam path which condensed it before it entered the eye of the viewer. However it should be pointed out that although this effect could be brought about by a simple lens used in the three point alignment system, the proportion of power in the central spots of the Fresnel zone plate and coarse grating image is only 10 % or less of the energy falling on that imaging element which, in any case, may itself be only a proportion of that radiated by the laser.

7. ACCURACY OF PERFORMANCE

The accuracy with which measurements can be conducted using the instruments and techniques described is affected by a considerable number of factors, the relative importance of which will vary from one application to the next. In particular, degradation of performance due to atmospheric effects is often difficult to predict other than in a rather general fashion, and a practical on-site test will often be the most reliable way of assessing the situation.

However, NPL experience of operating the three point system has shown that with very steady image conditions fiducial settings on the image can be made by most observers to within 0·05 mm. When the image is shimmering agreement between observers as to its mean position will often be within one twentieth of the amplitude of image movement. The possibility that this is superimposed on a long-term displacement due to atmospheric refraction has been stressed already. References 3, 4 and 10 quote accuracy values achieved in various practical applications of the three point system. For example, horizontal measurement of the deflection of the crest of a dam 360 m long with reference to end datums is being measured with an accuracy of ± 0.5 mm. A small battery-powered set of equipment for measuring out-of-flatness of steel plates forming part of box girder bridges with visual assessment of image position achieved an accuracy of ± 0.15 mm in 6 m. Equipment made to assist construction of very large overhead cranes having spans up to 60 m is expected to align the axles to within 0.15 mm.

As regards proprietary equipment, manufacturers' literature will usually provide some guide to performance, although note should be taken as to whether figures quoted make any allowance for degradation in a practical situation or whether they represent performance in idealised atmospheric conditions.

REFERENCES

1. *Encyclopaedia Britannica*, 11th edn, **25–26**, 557–72.
2. Dagnall, R. H. and Pearn, B. S. *Optical Alignment*, 1967, Hutchinson, London.
3. Harrison, P. W., Tolmon, F. R. and New, B. M. *Proc. Inst. CE*, **52** (1972), 1–24.
4. Harrison, P. W. *Civ. Eng. Public Works Rev.*, **68** (1973), 224–7.
5. Harrison, P. W. *Survg. Techn.*, **6** (1978), 6–7.
6. Hewlett-Packard. Measurement of straightness of travel. *Laser Measurement System Application Note 156-5*, 1976.
7. Tolmon, F. R. *Proc. Conf. Nelex*, **76** (1976).
8. Harrison, P. W. *Wat. Pwr. Dam Constr.*, **30** (1978), 52.
9. Van Milaan, A., *TNO-Nieuws*, **19** (1964), 471–6.
10. Bennett, S. J. and Gates, J. W. C., *J. Phys. E.*, **3** (1970), 65–8.
11. New, B. M. *Appl. Opt.*, **13** (1974), 937–41.
12. Cooper, M. A. R. *Modern Theodolites and Levels*, 1971, Crosby Lockwood, St. Albans, UK.
13. Harrison, P. W. *Proc. Conf. Laser 78*, 1978, Publ. Engineers Digest, London.

Chapter 5

LENGTH AND DISPLACEMENT MEASUREMENT BY LASER INTERFEROMETRY

S. J. Bennett

National Physical Laboratory, Teddington, UK

1. INTRODUCTION

The first measurements of length by interferometry were made by Michelson and Benoit at the end of the 19th century.[1] The primary metric standard of length at that time was the International Prototype Metre, a bar of platinum–iridium alloy inscribed with two fine lines 1 m apart. Michelson and his colleague measured the separation of this pair of lines in terms of the wavelength of the red line from a cadmium lamp at 644 nm. These measurements were repeated in 1905 by Benoit *et al.*[2] who achieved a result with an uncertainty of about 1 part in 10^7.

Subsequent measurements at a number of standards laboratories indicated that high precision could be achieved with interferometry[3] and the International Committee of Weights and Measures therefore agreed to a proposal that the metre standard bar should be replaced by a wavelength definition. The best available spectral line was the orange line corresponding to the transition between the levels $2p^{10}$ and $5d^5$ of the krypton-86 atom, and the definition of the metre as 1 650 763·73 wavelengths of this radiation was internationally accepted in 1960.[4]

The limited intensity and coherence of available thermal sources meant that the interferometric measurement of length and distance was still difficult and such measurements could only be made over distances of a few centimetres. The development of lasers has largely removed these limitations. The power output of a single-frequency helium–neon laser may be several thousand times the usable power from a standard krypton lamp, and the spectral purity of laser sources allows interference to be observed over distances up to several hundred metres in normal atmospheric conditions. Recent international agreement on a value for the wavelength of a laser stabilised to high precision by saturated absorption in iodine vapour has confirmed the potential accuracy of lasers as length standards.

Laser interferometry is a fast, accurate measurement technique which has been widely developed and applied to many different metrology problems. Interferometers are capable of operating in adverse conditions of vibration and atmospheric disturbance, although they are always susceptible to interruption of the optical paths. Needing no reference to a material standard of length, the laser interferometer is being used for the measurement of length and distance in industrial environments as well as in standards laboratories.

2. LASERS

The first helium–neon gas discharge laser was announced by Javan et al.[5] in 1961, and the following year White and Rigden[6] reported the first continuous visible output. Although helium–neon lasers can operate at several wavelengths in the visible and infra-red regions of the spectrum[7,8] and there are now many other gas laser systems available,[9,10] it is the original 633 nm wavelength which is almost exclusively developed and used for metrology.

The helium–neon laser is a relatively simple system requiring no special cooling, which can be operated from a normal single-phase power supply. Three hundred and sixteen nanometres (half a wavelength) is a conveniently short unit for precise dimensional measurement, and higher resolution can be obtained when necessary by subdividing this unit. The alignment of an interferometer is much simpler when the beams can be seen than with, for example, infra-red light, and conventional silicon photodiodes have adequate sensitivity to red wavelengths.

A gas laser is an active resonant cavity which may oscillate in one or more longitudinal modes simultaneously. The wavelength range over which a helium–neon laser may operate is about 3×10^{-6} of the mean wavelength. The simplest 1–2 mW lasers normally emit two distinct wavelengths separated by about 1×10^{-6}, and a laser of this type is an adequate source for interferometry over distances up to about 200 mm with an accuracy of a few parts in 10^6. For more accurate measurements, or for interferometry over longer distances, a stabilised laser emitting only a single wavelength is necessary. The first such lasers were only about 150 mm long and were stabilised by locking a single mode to the dip which is observed at the centre of the neon gain profile.[11−14] The stability of these 'Lamb-dip' lasers is generally a few parts in 10^8 and comparisons at several national laboratories[15] have demonstrated that they have an overall wavelength uncertainty of about 1 part in 10^7. Recently developed saturation techniques[16,17] in which a laser wavelength is locked to a particular hyperfine component in the molecular spectrum of iodine vapour have produced lasers with stabilities approaching 1 part in 10^{12}. This level of accuracy is not likely to be required in dimensional metrology but may have implications for the way in which the metre will be defined in the future.

Other stabilisation methods have been developed for laser inter-ferometry, and most of these generate a single output wavelength with a stability of about 1 part in 10^7. In addition to techniques making use of the Zeeman effect[18,19] there are also methods for stabilising general purpose 1–2 mW lasers with a simple thermal servo.[20,21] The two modes emitted by such a laser are generally orthogonally polarised so that a single wavelength can easily be isolated in the output.

Almost any correctly made stabilised helium–neon laser will have a stability of a few parts in 10^7 or better and the natural bandwidth of the laser will limit the maximum possible excursion from the means wavelength to 1·5 parts in 10^6. It is possible, however, that errors in manufacture could produce greater wavelength differences, so that a laser should be checked by comparison with a laser of known wavelength to ensure complete measurement confidence. Once checked it may be relied on as a wavelength standard within the above limits for the life of its discharge tube.

3.   INTERFERENCE

Light propagates as a wave motion and it undergoes reflection, refraction and diffraction (see Appendix, p. 267). The disturbance produced by a wave at a point in space can be described by its amplitude and phase. Thus, for a wave with a frequency f, this disturbance is given by

$$A = A_0 \cos (2\pi f t + \phi)$$

where A_0 is the amplitude and ϕ the phase at $t = 0$. The intensity of the light is proportional to the squared modulus of the amplitude (A_0^2) and the wavelength is $\lambda = c/f$ where c is the velocity of light.

If two beams of light have the same frequency (wavelength) and constant relative phase, they are said to be coherent with each other. When they illuminate a common area the superposition of the two leads to a resultant disturbance

$$A = A_1 + A_2$$
$$= A_{01} \cos (2\pi ft + \phi_1) + A_{02} \cos (2\pi ft + \phi_2)$$

Ignoring the amplitude difference and using the expression for the sum of two cosines leads to

$$A = A'_0 \cos (2\pi ft + \tfrac{1}{2}\overline{\phi_1 + \phi_2}) \cos \tfrac{1}{2}\overline{\phi_1 - \phi_2}$$

so that the resultant intensity depends on the relative phase $(\phi_1 - \phi_2)$ of the two waves. This effect is known as interference, and it is this which is responsible for the dark and bright bands in Young's experiment as well as for the coloured effects in an oil film on water. When two waves are in phase they reinforce each other to produce a bright interference maximum $(\phi_1 = \phi_2)$, but when in antiphase $(\phi_2 = \phi_1 \pm \mu)$ they cancel at a dark minimum.

In an interferometer, the phase difference between two beams at a point where they interfere is a simple function of the lengths of the optical paths from the source. The observed interference pattern then depends on the spatial distribution of the difference of path length and may take the form of straight bands, rings or more complicated patterns of interference fringes, which form contours of equal path difference. The word 'fringe', which normally refers to a border or margin, is possibly a slightly unfortunate one but the word is now firmly established in the language of optics.

In a simple Michelson interferometer (shown diagrammatically in Fig. 1) light from a source S is divided at a beam-splitter plate B. This reflects part of the light to a plane mirror M_1, while allowing some to be transmitted to M_2. The light reflected normally at the mirrors returns to the beam-splitter. Some of it then returns towards the source, while the remainder can interfere on a screen placed at A. The path difference (p) between the two beams at a point P on the screen is given by

$$p = SQR_2 QP - SQR_1 QP = 2(QR_2 - QR_1)$$

plus a small term taking account of non-geometrical path differences, e.g. the effects of changes of phase on reflection at the beam-splitter and mirrors. If the beams arriving at A have the same divergence, but are

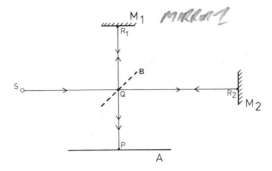

Fig. 1. Michelson interferometer.

slightly inclined to each other, then straight parallel fringes are observed. When the relative tilt is removed, but the beams diverge at different rates, the fringes will be in the form of concentric circles. However, when the beams are brought into coincidence by careful adjustment of the interferometer so that the path difference is everywhere the same with a fraction of a wavelength, then the fringes are said to be fluffed out to yield a uniform patch of light. The intensity of this patch varies sinusoidally as the relative phase of the two beams alters. This is the configuration which is generally sought in laser interferometry, and variations of the Michelson interferometer in which it is realised are described below.

4.   LASER INTERFEROMETRY FOR LENGTH MEASUREMENT

Laser interferometry is now widely used for length metrology over distances from micrometres to metres. In a laser interferometer the path difference is varied by moving one of the reflectors and the displacement is monitored by counting the cycles of interference. Many interferometer arrangements have been described[22,23] and four simple versions are shown in Fig. 2.

Figure 2(a) shows the simple Michelson interferometer of Fig. 1 illuminated from a laser source. If the plane mirrors are carefully adjusted, the two emerging beams are brought into coincidence and the interference fringes are fluffed out. The intensity on the detector, D, then varies through one cycle from bright to dark to bright again as one of the mirrors is displaced normal to its surface by one half wavelength of the laser light,

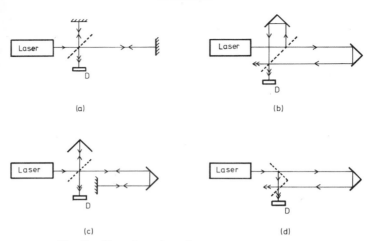

(a)

(b)

(c)

(d)

Fig. 2.   Four laser interferometer arrangements.

changing the path difference by one wavelength. There are two difficulties associated with this arrangement. Firstly, it is necessary to maintain the alignment of the mirrors to a very high accuracy, and secondly the light which returns into the laser may disturb it, causing fluctuations of intensity and wavelength. Both these problems may be overcome with the arrangement of Fig. 2(b), in which each of the plane mirrors is replaced with a cube-corner reflector.[24] This reflector consists of three mutually perpendicular plane reflecting faces and is usually made in the form of a single piece of glass which may be imagined to have been cut from the corner of a cube. The reflected beam emerges in a direction parallel to the incident beam though generally displaced from it, so that the alignment of the interferometer is not affected by tilting the reflectors. As the reflected beams are laterally displaced no light returns to the laser.

Although it is probably the most widely used arrangement that shown in Fig. 2(b) is susceptible to transverse displacements of the reflectors. An interferometer which is immune to tilt and transverse displacement of the moving reflector[25] is illustrated in Fig. 2(c), but this scheme has not found wide application. Figure 2(d) shows a fourth version of the interferometer which requires only one reflector, but also includes two beam-splitters which must be carefully adjusted to bring the two interfering beams into coincidence.

Each of the interferometers shown in Fig. 2 may be used to measure displacement by attaching the right-hand reflector to the component to be measured. If the component moves in one direction at a constant velocity, v, a sinusoidal signal is obtained from the detector at a frequency $f = v/\lambda$ where λ is the wavelength of the light. This signal can be counted to determine the distance travelled providing the moving reflector does not stop or reverse its direction of travel. In order to determine the direction and resolve the ambiguity when it reverses, a second output in phase quadrature to the first is required. The arrangements of Figs 2(b) and (d) are particularly convenient in this respect as they already have two outputs which can be brought into quadrature if a suitable metallic film is used for the combining beam-splitter.[26,27]

A complete measurement system, based on Fig. 2(b), is shown in Fig. 3. The laser, beam-splitter, detectors and one cube-corner reflector form the fixed part of the interferometer, while the second cube-corner is attached to the moving part. The outputs from the two detectors, which are usually silicon photodiodes, are counted in a bidirectional counter, one count of which corresponds to a displacement of the moving part of half a wavelength. The accumulated count may be displayed directly or converted to convenient units of length by multiplying by half the laser wavelength. If improved resolution is required additional electronic logic may be incorporated to divide each half wavelength into a number of parts.

The beam emitted from a laser is generally circular, with a gaussian intensity distribution which is a maximum at the centre and has a value

$$I_r = I_0 \exp\left(-2r^2/w^2\right)$$

at a distance r from the centre (see Appendix, p. 267). The constant w represents the radius at which the intensity is $1/e^2$ of its maximum value and is sometimes referred to as the spot size.[28] A circular area of diameter $3w$

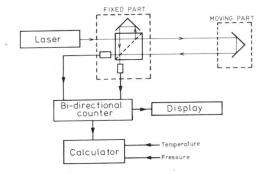

Fig. 3. Laser interferometer measurement system.

contains 99 % of the beam energy and corresponds approximately to the apparent beam size.

If a beam has a minimum spot size w_0, referred to as the beam waist, then the beam diverges from the waist and at a distance, z, the spot size, $|w_z|$, is given[29,30] by

$$w_z^2 = w_0^2 + (\lambda z/\pi w_0)^2$$

Thus, for a helium–neon laser at a wavelength of 633 nm, if $w = 1$ mm then $w_z = 1\cdot4$ mm when $z = 5$ m. This difference of diameter would lead to a loss of signal of about 6 % which would not be sufficient to cause the counter to miscount. An incident beam of diameter $(3w)$ of at least 4·2 mm is therefore required for a 5 m interferometer. Similarly, in a longer interferometer a beam with a spot size at the waist of 3 mm will diverge at 50 m to a spot size of 4·4 mm. This, too, would result in a 6 % loss of signal due to mismatch of beam size so that a 13 mm beam would be necessary for an interferometer of this length.

In addition to the effect of beam divergence, other factors may reduce the amplitude and level of the signals from the photodetectors of an interferometer. In particular, transverse displacement of the cube-corner reflector will produce a relative shear of the two interfering beams at the detector. If the two spots, of sizes w_1 and w_2, have their centres separated by a distance d the effect of the shear is to reduce the signal by a factor $\exp(-d^2/(w_1^2 + w_2^2))$.[23] For spots of equal size a shear of $0\cdot3w$ produces a loss of 4 %, which rises to 16 % at a shear of $0\cdot6w$.

5. POLARISING INTERFEROMETERS

Interferometers incorporating polarising beam-splitters have particular properties and advantages for many applications. A polarising beam-splitter generally consists of a series of dielectric layers of alternately high and low refractive index deposited on the diagonal face of a split cube.[31,32] A combination of a dozen layers can have a reflectivity greater than 99 % for the polarisation component normal to the plane of incidence and less than 1 % for the orthogonal component.

Figure 4 illustrates the properties of the Michelson interferometer of Fig. 1 with a polarising beam-splitter. The light from the source contains components polarised parallel and normal to the plane of the figure (i.e. it may be unpolarised or polarised at 45°). This is indicated by the symbols on the incident ray. The beams reflected and transmitted by the beam-splitter

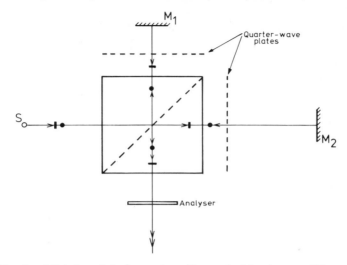

Fig. 4. Michelson interferometer with a polarising beam-splitter.

have mutually orthogonal polarisations, and in the absence of the quarter-wave retardation plates, placed as shown in each arm of the interferometer, the beams would both be returned in the direction of the source. The plates rotate the planes of polarisation through 90° so that the reflected beam, on returning to the beam-splitter, is transmitted. The transmitted beam is likewise reflected so that no light returns to the source. This is an important consideration if the amount of light available is small or the source is a stabilised laser which may be sensitive to retroreflected light.

The polarising interferometer may also be used with cube-corner reflectors, but when used with plane mirrors it can be rendered insensitive to tilt of the mirrors by double-passing each arm of the interferometer and incorporating a means of inverting the beams between passes.[33] The double-passing can be achieved with a polarising beam-splitter and quarter-wave plates as shown in Fig. 5, where the inverting component is a cube-corner reflector.

As in Fig. 4, the incident light, polarised at 45° to the plane of incidence, is divided at the polarising beam-splitter to produce two orthogonally polarised beams, shown slightly separated for clarity. Each beam passes through a quarter-wave plate before being reflected normally at one of the interferometer mirrors. As in Fig. 4 again, the plates are orientated so that the planes of polarisation of the beams returning to the beam-splitter have

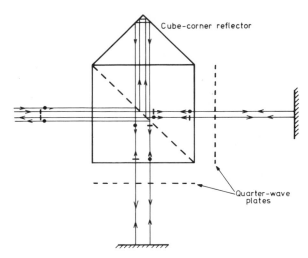

Fig. 5.   Double passed Michelson interferometer.

been rotated through 90°. The beam that was first transmitted is therefore now reflected and vice versa so that both beams now enter the cube-corner. This inverts the beams (reverses them about two orthogonal axes) and returns them to the same mirrors as before. Two more passes through the quarter-wave plates having restored the initial polarisation conditions, both beams leave the interferometer in a direction parallel to the incident beam. If an analyser is placed in the output beams interference fringes will be observed.

   If this interferometer is accurately manufactured, the interfering beams will always be mutually parallel so that the interference pattern will be 'fluffed out'. This parallelism is independent of the mirror alignment and the mirrors have only to be adjusted so that the beams overlap at the output.

   A useful variation of the arrangement of Fig. 5 is illustrated in Fig. 6, where a polarising interferometer is applied to the measurement of thermal expansion.[34] One mirror of the interferometer is formed by the top surface of the specimen under investigation which stands on a base plate within an evacuated oven. The base plate forms the second mirror of the interferometer which monitors the movement of the specimen top surface relative to the plate.

   The two beams produced at beam-splitter A are reflected at the polarising beam-splitter B and pass into the oven through a window. One of

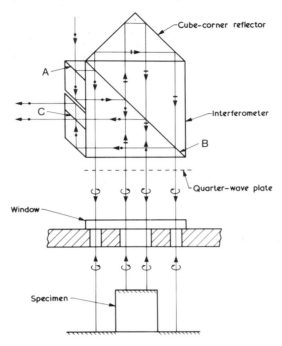

Fig. 6.   Interferometer system for thermal expansion measurement.

the beams is reflected normally at the upper face of the specimen, the other at the base plate. The quarter-wave plate once again rotates the plane of polarisation of the returning beams which are therefore transmitted by beam-splitter B. The cube-corner returns the beams to the oven, and after this second pass they are reflected at B to recombine at beam-splitter C. This third beam-splitter is a metal film which introduces a phase difference between the two outputs for bidirectional counting.[29]   The complete interferometer head of Fig. 6 can be manufactured as a single cemented unit. Extremely precise adjustment of the instrument is not then necessary and an adequate interference signal is easily obtained.

An alternative arrangement of the polarising interferometer can be used with a laser emitting light at two slightly different frequencies (f_1 and f_2) which have orthogonal polarisations.[35] In the case of a Zeeman split laser a frequency difference of about 2 MHz (4 parts in 10^9) is produced by placing the laser tube in a longitudinal magnetic field. The two components are separated at a polarising beam-splitter (Fig. 7) and directed to the two

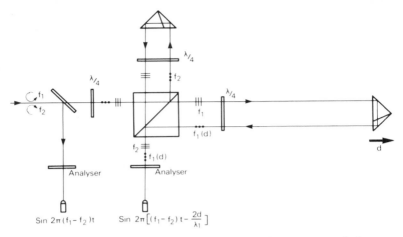

Fig. 7. Two-frequency interferometer for use with Zeeman split laser.

paths of the interferometer. The beams recombine as shown and pass through an analyser to a detector. This detector generates a beat signal at the difference frequency

$$S' = S'_0 \sin 2\pi t ((f_1 - f_2)t - 2d/\lambda)$$

which is similar to the beat signal obtained directly from the laser output

$$S = S_0 \sin 2\pi (f_1 - f_2)t$$

but with a phase which depends on the position of the moving cube-corner reflector. A counter which counts the difference between the two signals S' and S therefore accumulates one extra count as the reflector is displaced through a distance of half a wavelength.

This interferometer, which has been commercially developed as a measurement system, has the advantage that the detectors are always generating an a.c. signal, even when the reflector is stationary. They can be a.c. coupled to the following amplifiers, giving considerable immunity to misalignment and partial obstruction of the laser beam.

This immunity to factors which reduce the amplitude of the detected signal can also be obtained with the interferometer of Fig. 8. In this system, described by Downs and Raine,[36] four signals are generated instead of the normal two. Each of the outputs from the main beam-splitter is divided into two orthogonally polarised components and the four resultant signals are equally spaced in phase by $\pi/4$. Using three of these signals and

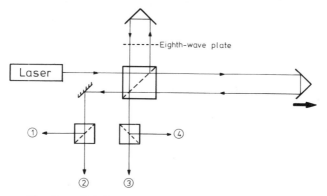

Fig. 8.　Michelson interferometer with four outputs.

subtracting various pairs it is possible to generate two signals in phase quadrature which are independent of changes of level at the detectors. In addition, the application of an automatic gain control improves further the tolerance of this system to signal loss, so that an interferometer system can be constructed which will tolerate a 90 % loss of signal due to misalignment, partial obstruction of the beam or reduction of source intensity.

6.　FLATNESS AND ALIGNMENT

A modified interferometer arrangement may be used to measure the flatness of a table or other machined surface or to monitor the pitch and yaw of a moving carriage. Two cube-corners are mounted as shown in Fig. 9 in a common block which may be attached to a carriage or free standing on a surface. Any tilt of the block about an axis normal to the plane of the figure will produce a change of path difference which is related to the angle through which the block turns and the separation of the reflectors. With a reflector separation of 65·3 mm, a tilt of 1 arcsec results in a path difference change of 1 fringe. When the double reflector unit is used for flatness measurement, one fringe corresponds to a difference of height of 0·5 μm if the spacing of the feet is 103·1 mm.

A useful alignment interferometer, in which the beam-splitter is a Wollaston prism, is illustrated in Fig. 10. The prism consists of two differently orientated quartz prisms cemented together so that the composite refractive index is different for two orthogonal polarisations

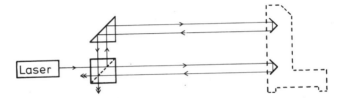

Fig. 9.   Double cube-corner flatness interferometer.

which emerge from the prism with a small angle between them. A compound mirror assembly, which consists of two plane reflecting surfaces mounted at an angle matching the angle between the beams, reflects the two beams which are then recombined in the Wollaston prism. Relative lateral displacement between the prism and the mirrors changes the optical path difference between the two beams. .

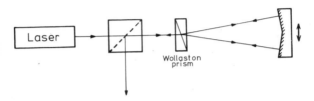

Fig. 10.   Alignment interferometer with Wollaston prism.

A lateral translation, x, introduces a change of path difference $2x \sin \theta/2$, where θ is the included angle between the beams. For $\theta = 1°$ the factor $2 \sin \theta/2$ is equal to $1/57$, so that 57 half-wavelengths of translation (18 μm) produce a single fringe count in the interferometer. The device is insensitive to deviations of the laser beam, and records only the required relative motion between the Wollaston prism and the mirror assembly. Small pitch and yaw motions do not create a path difference either and therefore do not affect the measurement accuracy of the interferometer.

7.   REFRACTIVE INDEX OF AIR

The refractive index of the air in the interferometer paths must be taken into account in calculating the laser wavelength. The velocity of light in air and therefore its wavelength is less than in a vacuum. If the refractive index of air is n, then the air wavelength is

$$\lambda_{\text{air}} = \lambda_{\text{vac}}/n$$

where λ_{vac} is the vacuum wavelength. Edlen[37] has published equations from which the refractive index of air at a particular wavelength can be calculated if the temperature, total atmospheric pressure and water vapour partial pressure are known.

The dependence of the refractive index on temperature for three different values of the pressure at 50 % relative humidity is shown in Fig. 11. It can be seen that the refractive index increases by approximately one part in a million when the temperature falls by one degree centigrade or the pressure increases by 400 pascals (3 torr). Quite a large variation of relative humidity ($\sim 50 \%$) is required to change the refractive index by a similar amount. Thus the accuracy with which these atmospheric parameters must be determined to obtain a given overall measurement accuracy may be estimated, and the data supplied to the calculator as shown in Fig. 3 either manually or directly from suitable atmospheric transducers.

It is immediately evident that a laser interferometer is rather sensitive to atmospheric changes, and care is necessary to obtain accuracies

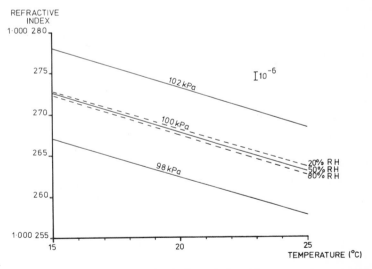

Fig. 11.   Variation of the refractive index of air with temperature at different values of the atmospheric pressure and relative humidity (RH).

approaching one part in a million, particularly in a hostile or disturbed environment or during stormy or changeable weather. It must also be borne in mind that interferometric measurements are in effect made with a gauge which has a coefficient of thermal expansion of 1×10^{-6} per degree centigrade, very different from that of, for example, steel. This means that measurements on machines or measuring instruments in an environment where the temperature is poorly stabilised may be more difficult to perform accurately and confidently than by more traditional methods of verification with steel gauges. The inherent resolution and accuracy of the laser interferometer can only be fully realised in practice if sufficient attention is paid to temperature (and pressure) measurement.

8. LASER INTERFEROMETRY IN THE STANDARDS LABORATORY

Because of the potentially high accuracy of the technique and the stability and reproducibility of laser wavelengths, laser interferometry is now widely used in standards laboratories for the calibration of gauges and other material length standards. Many interferometer systems have been described[38-40] and at the National Physical Laboratory (NPL) line standards, metrological gratings, length bars, geodetic tapes and thread gauges are all calibrated directly by laser interferometry. To minimise the effects of geometrical errors and misalignments, each interferometer forms an integral part of the instrument to which it is attached.

The NPL's automatic scale measuring interferometer[41] measures automatically the positional errors of lines on metrological scales and gratings up to 1 m in length. A carriage, bearing the scale, moves continuously along the bed of the machine (Fig. 12), passing beneath a photoelectric microscope which detects the positions of the scale lines and generates a pulse each time a line passes. To calibrate a grating, the microscope is replaced with a grating reading head.

Two mirrors at the left-hand end of the carriage, as seen in the photograph, form the moving reflector of an interferometer system similar to that of Fig. 2(d), folding the laser beam around the scale so that the measurement axis and the scale axis coincide. A single detector receives the sinusoidally varying intensity as the carriage moves along the machine, and the cycles of interference are counted electronically. One cycle (fringe) corresponds to a carriage motion of half a wavelength (about $0.316\,\mu$m) and higher resolution is obtained by interpolating electronically to about

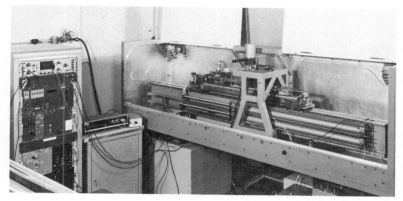

Fig. 12. NPL automatic scale measuring interferometer.

1/50 of a fringe. This is achieved by assuming that the carriage velocity is constant over a short distance and timing the interval between the rising edge of a photomultiplier pulse and the subsequent pulse from the interferometer electronics. The time between this pulse and the next is also measured and the required fraction is obtained as the ratio of the two time intervals.

The outputs of the main fringe counters and the timers are recorded on punched tape for each scale line, together with details of the scale temperature and atmospheric conditions. The bar is passed in both directions under the microscope by reversing the direction of the carriage motion in order to record the positions of both edges of each scale line and all the data are reduced by a computer program which calculates the positional errors of the lines. The accuracy of determination of line position on a good line standard is typically $\pm 0.3\,\mu$m.

The ability of laser interferometry to make measurements over long distances is demonstrated by the NPL 50 m interferometer for geodetic tapes.[42] To remove the need for the long and painstaking series of measurements required to establish a 50 m base line, by stepping off the distance with a 4 m standard, an interferometer was installed on the NPL comparator bench.

The bench itself forms a stable base[43-45] and the interferometer carriages run on 25·4 mm diameter rails along the top of the bench. These rails have been carefully aligned so that the maximum deviation from a straight line of a point on a carriage running on the rails is less than 0·5 mm along the 50 m length.

The arrangement of the interferometer, which is a folded version of the double cube-corner interferometer of Fig. 2(b), is shown diagrammatically in Fig. 13. The interferometer components are mounted on two carriages, each of which also bears a visual microscope for observing the graduation lines on a tape under calibration. A fixed platform at one end of the bench supports the single-frequency helium–neon laser and two PIN photodiodes which detect the interferometer signals.

A tape under calibration is supported in multiple catenary along the front of the bench and the expanded beam from the laser is directed onto a horizontal axis behind the tape. This beam is divided at a beam-splitter on the right-hand carriage and the reflected beam bridges the tape via a cube-corner reflector to a second beam-splitter in front of the tape. This beam thus forms the fixed arm of the interferometer, while the beam transmitted at the first carriage forms the variable arm, continuing to second (left-hand) carriage. This is similar to the first, except that it has two full mirrors instead of beam-splitters. The second beam is therefore also folded over the tape, and returns to combine with the first at the right-hand carriage. As in the case of the scale-measuring interferometer, the optical paths are arranged in such a way that the measuring plane contains the tape graduations which lie on the measurement axis.

The interference signals are detected at the two PIN photodiodes, the outputs of which are amplified and counted in an electronic bidirectional counter. Because the level of light intensity at the detectors may vary as the carriages traverse over large distances, the signal amplifiers incorporate automatic compensation which adjusts the level while the carriages are in motion and remembers it when they are stationary.[46]

It will be seen that the interferometer monitors the separation of the two carriages. As the microscopes are mounted on them, measurements can be

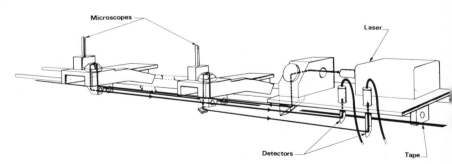

Fig. 13.   NPL 50 m interferometer for surveying tapes.

made directly on a tape without establishing a fixed base. An ordered sequence of observations by two observers enables an overall uncertainty of one part in 10^6 to be achieved with this instrument, and its performance has been verified by reference to tapes with a well established calibration history.

Figure 14 is a photograph of a commercial laser interferometer system on a measuring machine for investigating the pitch errors of tapered screw thread gauges for the petroleum industry. The laser, in the centre of the picture, is mounted on the main casting of the instrument and the whole of the casting can be tilted to align with the taper of the gauge being measured (front right). The cube-corner reflector, which is visible in its cylindrical enclosure half way up the vertical column at right, is mounted at the top of a tapered arm, the lower end of which is fitted with a stylus which can be located in the thread of the gauge. The arm is biased so that the stylus is pressed into each thread in turn while the complete stylus/cube-corner assembly moves freely on two lapped vertical columns. The interferometer, which is of the type shown in Fig. 7, records the movement of the reflector, so that the variation in pitch from one thread to the next can be

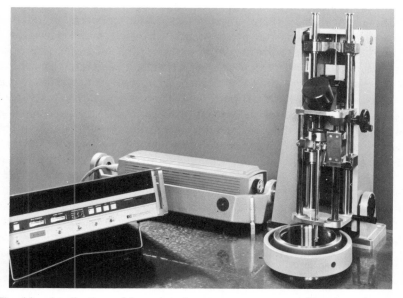

Fig. 14. Application of laser interferometer system to calibration of a screw thread gauge.

investigated. The display box at the left of the photograph displays the distance moved by the reflector in quarter-wavelength units or in millimetres or inches using a conversion factor for the refractive index of air, entered on thumbwheel switches.

9. OTHER APPLICATIONS

In addition to the measurement of gauges and length standards as described above, laser interferometry has many other scientific and industrial applications where high accuracy is required. In particular, there are a large number of physical measurements which depend on the determination of length, change of length or displacement.

Some of the most sensitive dimensional measurements are those of earth strain. These have been particularly important for the observation and investigation of earth tides and earthquakes. Several laser interferometer strainmeters have been reported, at least two of which exploit the exceptional wavelength stability of lasers stabilised by saturated absorption.

Poorman's Relief Mine in Boulder County, Colorado, USA is the site for a 30 m spherical Fabry-Perot interferometer built by Levine and Hall.[47] The two interferometer mirrors are mounted on piers embedded in the bedrock and the entire instrument is enclosed in a tube evacuated to a pressure of 1·3 Pa (10 mtorr). The interferometer is illuminated with a helium–neon laser operating at 3·39 μm. One of the mirrors of this laser is mounted on a piezo-electric support so that the wavelength can be tuned over a small range. This wavelength is servo-controlled to a transmission maximum of the interferometer and changes in the separation of the interferometer mirrors thus produce corresponding variations in the laser wavelength. These are monitored by mixing part of the output with the output from a second laser which is independently locked to a molecular absorption line in methane. This mixing is performed at a fast photodiode and the resulting beat frequency is recorded. Since changes in the length of the interferometer due to geophysical strain are reflected in the beat frequency signal, they can be monitored with a very high sensitivity. The drift of this instrument is less than 10^{-10} strain per day and the noise level is below 10^{-12} strain.

Goulty et al.[48] have described a similar strainmeter operating in Queensbury Tunnel, Yorkshire, UK. This is an unequal path Michelson interferometer operating over a distance of 54 m. The principle of operation is similar to that of the instrument of Levine and Hall, but alignment and operation are simplified by using an interferometer arrangement similar to Fig. 2(b) with a visible helium–neon laser.

The acceleration due to gravity can be determined by recording the motion of a freely falling body. Hammond and Faller[49] used a cube-corner reflector as the falling body and measured the gravitational acceleration with an accuracy better than 5 parts in 10^8. The gravity-meter of Sakuma,[50] at the International Bureau of Weights and Measures, uses laser interferometry to measure the acceleration of a cube-corner in free rise and fall. With an accuracy better than 1 in 10^8, this instrument has monitored earth tides and other geophysical variations, as well as the secular variations of gravity.

The mercury U-tube manometer is still widely used as a pressure standard in the atmospheric range. The measurement of pressure with such an instrument involves a knowledge of the density of mercury, the acceleration due to gravity and the differential height of the mercury column. The first two may, under favourable circumstances, be known with an uncertainty of less than 1 part in 10^6 and a measurement of the height difference must be made with comparable accuracy. Interferometric methods which use the mercury menisci as interferometer mirrors are generally frustrated by disturbances of the liquid surfaces. Tilford[51] has overcome this problem by using the 10·6 μm wavelength radiation from a CO_2 laser to monitor the motion of the mercury at speeds up to 5 mm min^{-1}. An alternative solution[52] is to use the mercury surface as the mirror of a lens/mirror 'cat's eye' reflector for visible light. This combination is highly immune to surface disturbances, while permitting direct interferometric measurement of surface motion over a distance of 400 mm at speeds up to 1 mm s^{-1}.

A somewhat similar application to the measurements on a mercury manometer have involved interferometric measurements on the changes in the level of a reservoir.[53] Changes in level can be due to both leakage and evaporation and in order to separate the two effects very sensitive measurements must be made during the short periods when evaporation is negligible (high humidity and zero wind). An automatically recording Michelson interferometer was used, with reflection from the water surface being provided by a floating cube-corner. The lateral movement of this float, up to 1 mm or so, was made unimportant by expanding the laser beam by a × 10 telescope, and the float was kept in position within this tolerance by using a protective tube containing a central magnet around the float which, in turn, also carried a magnet at its centre. Recording was on scales of either 7·5 μm or 750 μm, the latter being more useful for the long-term (daily) changes.

In all the previous examples, the main interest has been with changes in position, or with displacement from one fixed point to another, the measurements being made with a mirror or cube-corner reflector which is translated between the two points. An alternative requirement is the precise measurement of distance between two points without the need for any travelling component other than the light beam. Such measurements have been made over large distances using time-of-flight measurements with pulsed laser or microwave sources, but these methods lack the high precision of interferometry. The problem had already been solved by Benoit in 1898, using several spectral lines and the method of exact fractions.[54] The principle involved the determination of the path difference in an interferometer by solving a set of simultaneous equations in the form $L = (n + \delta)\lambda/2$ for several different wavelengths (λ). Using the measured values of the 'fraction', δ, and trying successive values of the integers (n), the equations can be solved to give a value for the length, L.

This method has been used routinely for nearly 50 years for the interferometric calibration of gauge blocks. A Fizeau interferometer, designed at NPL and manufactured by Hilger and Watts, uses four wavelengths from a cadmium discharge lamp and is a practical instrument for industrial and laboratory use.[55] A batch of gauges can be measured by wringing them to a rotatable base plate so that each gauge in turn can be brought into the field of view for measurement. Gauges larger than 25 mm can be measured in this instrument with an uncertainty of one part in a million.

The addition of a laser source offers several improvements, notably in the intensity and contrast of the interference fringes as well as in the extension of the method to longer lengths. Several spectral lines from a CO_2 laser have been used[56] and the use of a dye laser[57] simplifies the subsequent calculations, as the wavelengths can be adjusted to produce any required values for the fractions (e.g. zero). A practical infrared multiple wavelength interferometer for length measurement has been built at the National Research Laboratory of Metrology in Tokyo.[58] This instrument uses a helium–xenon laser, operating at 3·37 and 3·51 μm, and can be applied to a range of metrology problems.

This chapter has concentrated on the description of laser interferometer systems as transducers for length measurement, but interferometry lends itself readily to the study of spatial problems in more than one dimension. The deviation from flatness of silicon wafers for semiconductor processing can be demonstrated by interferometric techniques in which the observed interference fringes are height contours of the silicon surface. In the past, the reduction of such fringe patterns to numerical results has been carried out by skilled operators, but this is expensive and the use of laser illumination has enabled automatic analysis of two-dimensional fringe patterns (interferograms) using digital television techniques or CCD cameras to scan the patterns. In this way the intensity variation of an interferogram along the scan lines is obtained in digital form, and these intensity data can be used in various ways. Parabolic curve fitting techniques have been used to locate the centres of fringes,[59] and the least-squares sine wave fitting routine described in Chapter 3 for moiré fringe analysis has also been used with two-beam interferograms for a similar purpose. These techniques are limited to simple linear fringe patterns (basically straight parallel fringes) and more generalised patterns usually involve a fair amount of interactive processing[60,61] which avoids the anomalous results that can occur with noisy patterns.

For phase measurements within a pattern of fringes, heterodyne techniques,[62,63] using two slightly different wavelengths to produce a lower beat frequency (as already described for the orthogonal polarisation interferometer) provide the greatest accuracy, and a combination of this method with a high speed scanning system offers a versatile means of analysing interferograms. Massie[64] describes a high speed image dissector camera controlled from a fairly modest microcomputer which also analyses the data. This system has a serial data acquisition rate of 50 μs per point, a resolution of 500 points per line and a phase accuracy of $\lambda/70$.

Laser interferometers have found many applications in physics, engineering and metrology, usually in one of the arrangements described in this chapter. High resolution and accuracy are obtained if sufficient care is taken to align the interferometer with the direction of motion[65] and to minimise thermal effects. In the generation and measurement of photomasks for integrated circuits, two-axis measurement systems are currently being used to good effect,[66] illustrating the ability of laser interferometry to meet the most demanding metrological requirements of modern production technology.

REFERENCES

1. Michelson, A. A. and Benoit, J. R. *Trav. et Mem. BIPM*, **11** (1895), 1–237.
2. Benoit, J. R., Fabry, C. and Perot, A. *Trav. et Mem. BIPM*, **15** (1913), 1–134.
3. Barrell, H. *Proc. Roy. Soc.*, **A186** (1946), 164–70.

4. Comptes Rendus. *11ème Conf. Gen. Poids et Mesures*, 1960, Gauthiers–Villars, Paris.
5. Javan, A., Bennett, W. R. and Herroitt, D. R. *Phys. Rev. Lett.*, **6** (1961), 106–10.
6. White, A. D. and Ridgen, J. D. *Proc. IRE*, **50** (1962), 1697.
7. Bloom, A. L. *Appl. Phys. Lett.*, **2** (1963), 101–2.
8. McFarlane, R. A., Bennett, W. R. and Lamb, W. E. *Appl. Phys. Lett.*, **2** (1963), 189–90.
9. Zory, P. *IEEE J. Quant. Elect.*, **QE-3** (1967), 390–8.
10. Goldsborough, J. P. *Appl. Phys. Lett.*, **15** (1969), 159–61.
11. Ballik, E. A. *Phys. Lett.*, **4** (1963), 173–6.
12. Rowley, W. R. C. and Wilson, D. C. *Nature*, **200** (1963).
13. Baird, K. M., Smith, D. S., Hanes, G. R. and Tsunekane, S. *Appl. Opt.*, **4** (1965), 569–71.
14. White, A. D. *IEEE J. Quant. Elect.*, **QE-1** (1965), 349–57.
15. Mielènz, K. D., Nefflen, K. F., Rowley, W. R. C., Wilson, D. C. and Engelhard, E. *Appl. Opt.*, **7** (1968), 289–93.
16. Hanes, G. R. and Dahlstrom, C. E. *Appl. Lett.*, **14** (1969), 362–4.
17. Wallard, A. J. *J. Phys. E.*, **6** (1973), 793–807.
18. Skolnick, M. L., Polanyi, T. G. and Tobias, I. *Phys. Lett.*, **19** (1965), 386–7.
19. Ferguson, J. B. and Morris, R. H. *Appl. Opt.*, **17** (1978), 2924–9.
20. Balhorn, R., Kunzmann, H. and Lebowsky, F. *Appl. Opt.*, **11** (1972), 742–4.
21. Bennett, S. J., Ward, R. E. and Wilson, D. C. *Appl. Opt.*, **12** (1973), 1406.
22. Rowley, W. R. C. and Wilson, D. C. *Proc. Inst. Mech. Eng.*, **183**, part 3D (1969), 29–33.
23. Rowley, W. R. C. *Alta Frequenza*, **41** (1972), 887–96.
24. Peck, E. R. *J. Opt. Soc. Am.*, **47** (1957), 250–2.
25. Terrien, J. *Revue d'Optique*, **38** (1959), 29–37.
26. Peck, E. R. and Obetz, S. W. *J. Opt. Soc. Am.*, **43** (1953), 505–9.
27. Raine, K. W. and Downs, M. J. *Optica Acta*, **25** (1978), 549–58.
28. Boyd, G. D. and Gordon, J. P. *Bell Syst. Tech. J.*, **40** (1961), 489–508.
29. Boyd, G. D. and Kogelnik, H. *Bell Syst. Tech. J.*, **41** (1962), 1347–69.
30. Kogelnik, H. and Li, T. *Appl. Opt.*, **5** (1966), 1550–67.
31. Heavens, O. *Optical Properties of Thin Films*, 1955, Butterworths, London.
32. Banning, M. *J. Opt. Soc. Am.*, **37** (1947), 792–7.
33. Bennett, S. J. *Opt. Comm.*, **4** (1972), 428–30.
34. Bennett, S. J. *J. Phys. E.*, **10** (1977), 525–30.
35. Dukes, J. N. and Gordon, G. B. *Hewlett Packard J.*, **21**(12) (1970), 2–8.
36. Downs, M. J. and Raine, K. W. *Precision Eng.*, **1** (1979), 85–8.
37. Edlen, B. *Metrologia*, **2** (1966), 71–80.
38. Koning, J. and Schellekens, P. H. J. *Annals of CIRP*, **19** (1971), 255–8.
39. Giacomo, P., Hamon, J. Hostache, J. and Carre, P. *Metrologia*, **8** (1972), 72–82.
40. Lenkova, G. A., Lokhmatov, A. I., Gurin, E. I., Koronkevich, V. P., Kolesova, E. B. and Tarasov, G. G. *Meas. Tech.*, **14** (1971), 1822–5.
41. Rowley, W. R. C. and Stanley, V. W. *Machine Shop.*, **26** (1965), 430–2.
42. Bennett, S. J. *Survey Review*, **22** (1974), 270–5.
43. Clark, J. S. *The Engineer*, **189** (1950), 201–3.

44. Clark, J. S. *The Engineer*, **189** (1950), 228–30.
45. Johnson, L. O. C. *The Engineer*, **203** (1957), 632–4.
46. Bennett, S. J. and Rowley, W. R. C. *J. Phys. E.*, **6** (1973), 963–4.
47. Levine, J. and Hall, J. L. *Geophys. Res.*, **77** (1972), 2595–609.
48. Goulty, N. R., King, G. C. P. and Wallard, A. J. *Geophys. J. Roy. Astr. Soc.*, **39** (1974), 269–82.
49. Hammond, J. A. and Faller, J. E. *IEEE J. Quant. Elect.*, **QE-3** (1967), 597–602.
50. Sakuma, A. *Special Publication 343*, 1971, US National Bureau of Standards, Boulder, Colorado, pp. 447–56.
51. Tilford, C. R. *Rev. Sci. Instrum.*, **44** (1973), 180–2.
52. Bennett, S. J., Clapham, P. B., Daborn, J. E. and Simpson, D. I. *J. Phys. E.*, **8** (1975), 5–7.
53. Jacobs, S. F. and Small, J. G. *Appl. Opt.*, **20** (1981), 3508–13.
54. Benoit, J. R. *J. Phys.*, **7**(3) (1898), 57.
55. Poole, S. P. and Dowell, J. H. Application of interferometry to the routine measurement of block gauges. In: *Optics in Metrology*, ed. P. Mollet, 1960, Pergamon Press, Oxford, pp. 405–19.
56. Bourdet, G. L. and Orszag, A. G. *Appl. Opt.*, **18** (1979), 225–30.
57. Bien, F., Camac, M., Caulfield, H. J. and Ezekiel, S. *Appl. Opt.*, **20** (1981), 400–3.
58. Matsumoto, H. *Appl. Opt.*, **20** (1981), 231–4.
59. Pugh, D. J. and Jackson, K. *Proceedings of Nelex Metrology Conference*, 1982, National Engineering Laboratory.
60. Funnell, W. R. J. *Appl. Opt.*, **20** (1981), 3245–50.
61. Robinson, D. W. *Proc. SPIE*, **376** (1983), to be published.
62. Massie, N. A. *Heterodyne Interferometry in Optical Interferograms— Reduction and Interpretation, ASTM STP 666*, eds. A. H. Guenther and D. H. Liebenberg, 1978, ASTM, Philadelphia.
63. Massie, N. A., Nelson, R. D. and Holly, S., *Appl. Opt.*, **18** (1979), 1797–803.
64. Massie, N. A. *Appl. Opt.*, **19** (1980), 154–60.
65. Hoffer, T. M. *Hewlett Packard Application Note 156–4*, pp. 27–32.
66. Koch, J. K. *Proc. SPIE: Developments in Semiconductor Microlithography*, **80** (1976), 112–20.

Chapter 6

HOLOGRAPHY AND ITS APPLICATIONS

E. R. ROBERTSON and W. KING

Department of Mechanics of Materials, University of Strathclyde, Glasgow, UK

1. INTRODUCTION

The name holography is derived from two Greek words *holos graphos* meaning, roughly, the 'whole picture' or record; the original, more prosaic term is wavefront reconstruction which provides a better technical description of what the process achieves, perhaps, but which conveys nothing of the sheer alchemy of a subject that has brought a little of the sense of wonder back into an age of scientific sophistication as well as opening up for the engineer and scientist a whole new range of optical techniques.

In 1948 Denis Gabor[1] published the results of his efforts to improve the resolving power of electron microscopes—he had invented holography. At that time, however, his efforts were considerably impeded by the lack of a satisfactory coherent source of light and it was not until the laser became available in the early 1960s that Leith and Upatnieks[2,3] were able to develop Gabor's invention to the point where it was possible to make fully three dimensional images of solid objects.[4] Since that time in 1964 the progress of holography has been spectacularly swift, the most important developments being those of holographic interferometry which had already begun by the end of 1964.

In this chapter an attempt will be made to help the reader to understand the basic principles and methods of holography and to indicate the more important areas of application of holographic interferometry. The emphasis will be on developing an intuitive feeling for the subject rather

161

than a full mathematical appreciation; for those wishing to make a more comprehensive study by more traditional analytical methods the references provided will indicate suitable sources.

2.   THE HOLOGRAPHIC PROCESS

Strictly speaking there are several holographic processes, most of them are optical but others employ acoustic, microwave and other techniques; however, as the most commonly used method is laser holography and as this is the only truly optical technique it is this which will be considered in this chapter.

To understand how it works, first consider how a single grain of photographic emulsion responds to light. It absorbs, or integrates all the light energy that lands on it during its exposure and blackens to a corresponding degree when processed. Examination of the grain afterwards will tell how much energy it has received but that is all. It cannot provide any information about the nature of the light, from which direction it arrived or about the phase relationship of the light arriving at the grain with that arriving anywhere else. So, if we take a source of light and illuminate an object for a period of time as in Fig. 1, so that some of the light is reflected back from the object onto a photographic plate and then we process the plate, all we get is a darkened plate! Even if the source of light is a laser we are no better off.

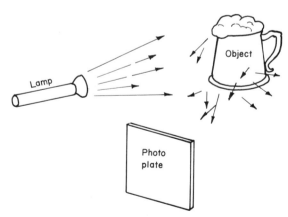

Fig. 1.   No matter how bright the prospect, this photographic plate will have a dark outlook.

Gabor realised that if he used a coherent source of light and divided a beam into two parts, illuminating the object with one part while directing the other towards the photographic plate, then the light reflected by the object onto the photographic plate would maintain a constant phase relationship with that going directly to the photo-plate (called the reference beam) from the source. This constant phase relationship would result in a complex standing wave pattern being formed at the photo-plate by the mutual interference of the two sets of light rays. A photo-plate exposed for a suitable period of time in such a situation would yield after processing a sort of coded message describing the object as seen from all possible parts of the photo-plate—a truly three-dimensional description if it could be decoded. Unfortunately, the light sources available to Gabor at that time were of such limited coherence that he could only cover objects of negligible depth—transparencies in the first instance. The invention of the laser in about 1960 introduced the coherent source Gabor had lacked.

Leith and Upatnieks developed Gabor's invention further and, using a laser as a light source, went on to produce the first holograms of solid, deep objects. The simplest way in which this can be done is shown in Fig. 2. The beam of laser light is expanded by a lens to illuminate both the object and a mirror placed alongside it. The object reflects scattered rays in all directions but some of them land on the photo-plate where all the grains of photographic emulsion can be considered to be looking at their own particular view of the object—this is called the object beam. The mirror is arranged to reflect the rays landing on it to the photo-plate. This is called the reference beam. These two sets of light rays landing on the photo-plate set up a complex pattern of interference effects which, because the two sets of rays are derived from the same source and are therefore mutually

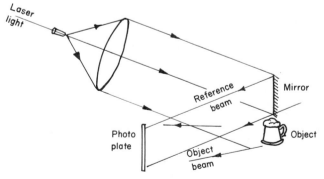

Fig. 2. The simplest set-up for reflection holography.

coherent, is constant with time—a standing wave pattern. After allowing a sufficient period of time for the photo emulsion to be adequately exposed the plate is developed, fixed, washed and dried. If it is now replaced in its original position and illuminated by the reference beam alone (e.g. by placing a piece of black card in front of the object so that the object beam is blocked off) then, on looking through the photo plate we see a virtual image of the object just as if the object was still in plain sight. By moving our position to look through different parts of the plate we see different perspective views of the object's image; it appears quite solid. What we are seeing, in fact, is not just a simple two-dimensional representation of the object such as a camera gives, nor even the limited three-dimensional view from a single position given by a stereoscopic pair but, rather, the 'whole picture'. What better name, then, for the processed plate than 'hologram'. When made this way a hologram looks like an unintelligible collection of finger prints when viewed in white light. A little thought will show the reader that these patterns can have nothing to do with the holographic process which is concerned with interference effects between light rays of about 500×10^{-9} to 600×10^{-9} m long. Such effects will be undetectable to the naked eye. Instead, the dirty appearance of the hologram results from the diffraction produced by dirt, etc., in the beam of the laser and other flaws in the system.

While it is easy to accept that a hologram will contain a complex description of an object the explanation which has been given of the reconstruction process has not explained *why* it works, only *how*. It is perfectly possible to use holography without really understanding it but that is hardly a satisfactory situation to be in. A concise mathematical description is given at the end of this chapter but an alternative, less rigorous approach is presented here which it is hoped will help the reader develop a feel for the subject.

In Fig. 3(a) two mutually coherent point sources of light S_1 and S_2 illuminate a surface some distance away where interference occurs between their rays. Suppose they arrive in exact antiphase (half a wave out of step); they will try to cancel each other out and so will create a dark fringe at this point. Then we can say that

$$(S_1A - S_2A) = (n + \tfrac{1}{2})\lambda$$

where λ is the wavelength of the light, and n is some integer. Moving along the surface, if the next location of destructive interference is at B then

$$(S_1B - S_2B) = (n + 1\tfrac{1}{2})\lambda$$

i.e.

$$(S_1B - S_2B) - (S_1A - S_2A) = \lambda$$

or

$$(S_1B - S_1A) - (S_2B - S_2A) = \lambda$$

i.e.

$$aB - bB = \lambda$$

or

$$d(\sin \theta - \sin \phi) = \lambda$$

$$d = \frac{\lambda}{\sin \theta - \sin \phi}$$

Hence, the smaller the difference between θ and ϕ the greater is d, the fringe spacing; that is, the closer together the point sources S_1 and S_2 the broader the fringe spacing—and vice versa. (Of course, Fig. 3(a) shows a grossly exaggerated diagram; in real terms the dimension d will be microscopically small compared with S_1A, etc.) Hence, in Fig. 3(b) if X and Y are points on an object's surface and R is a reference source then, on the photo-plate shown, at the location P fine fringes will be formed by the mutual interference of rays arriving from X and R and broad fringes will be formed by those from Y and R. In this way, fringes will be formed all over the photo-plate by the interference of rays arriving from all points on the object surface with those from the reference source, the fringe spacings being related to the relative positions of the contributing points.

It can be assumed that the reference source provides an even, consistent amplitude of light throughout its field of illumination; however, the rays derived from the object surface will vary in amplitude according to the reflectivity of the surface—dark parts reflecting weak rays, light parts stronger rays and shiny parts being almost completely spectral in their reflectance. Because of this the degree of interference (or depth of modulation) occurring at any point on the photo-plate will depend on the surface appearance of the part of the object contributing to that particular point. So, not only does the photo-plate contain a description of the location of all points of the object surface it also records their full light and shade appearance, so to speak.

Reconstruction of the image occurs by diffraction of the rays of the reference beam if a hologram is made as described above and then replaced

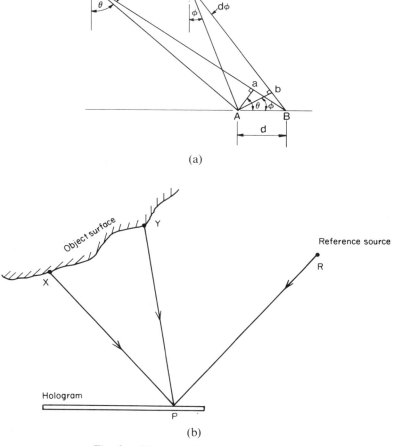

(a)

(b)

Fig. 3. The geometry of interference.

in the reference beam alone (or in a similar beam derived from a geometrically similar position as the reference was in the recording process). It will be recalled that when a beam of monochromatic light passes through a diffraction grating some of the light goes straight through (the zero order), some is bent through an angle (the first order), some is bent through a greater angle (the second order) and so on. Only the zero order and the first order usually show much strength and, indeed, by choosing an appropriate geometry, i.e. by having the reference beam derive from a fairly oblique angle, other orders may be denied an existence.

Considering the first order alone, for the present, it will also be recalled that the angle through which the first order is bent on diffraction depends on the grating fringe spacing, the angle increasing as the fringe spacing decreases, and vice versa. So it is that a reference beam ray is diffracted or bent on transmission through the hologram according to the fringe spacings recorded at the point through which it passes; these, in turn, depend on the location of the object point which contributed to that fringe spacing. Also, depending on the brightness of that object point and the degree of interference it created with the original reference beam, so an appropriate *amount* of the reference beam will be diffracted into the first order.

Hence, in this way, it is possible to reconstruct the complete wavefront reflected by the object in location, phase and amplitude. All of this is illustrated in Fig. A1 in the mathematical derivation in the appendix.

3. SOME PRACTICAL CONSIDERATIONS

Lasers can be broadly classified into two types for our present purposes, namely continuous wave (c.w.) and pulsed lasers. Usually, but not always, c.w. lasers are gas filled and pulsed lasers are solid state, such as ruby, etc., but the division is not as precise as that. As their names imply, a c.w. laser gives a constant output while it is switched on whereas a pulsed laser delivers its light in bursts or pulses, usually of a very short duration (nano- or even pico-seconds typically).

Whichever kind of laser is used its output must have adequate coherence for the kind of holography to be undertaken. In particular, the coherence length of the laser will define the object-volume that can be examined or holographed. For example, a helium–neon (HeNe) c.w. laser might have a coherence length of 25 cm; this means that the maximum permissible difference in the path lengths of the object beam and reference beam is 25 cm for ordinary holography. For path lengths greater than this the phase relationships are no longer sufficiently constant with time and so the holographic image will 'fuzz' out. In the case of a pulsed laser the coherence length may even be restricted by the pulse time itself.

The holographic process depends on the establishment of phase relationships between light waves arriving at the photo-plate which stay constant during the time necessary for adequate exposure of the plate. If these phase relationships change relatively by even as little as 0.1λ, say, during the time of exposure the consistency of the interference pattern established between them is severely threatened. Thus it can be seen that holography demands considerable positional stability in the positions of the various elements involved. For the majority of c.w. laser holography it is common to mount the equipment on a vibration isolated table and stringent precautions are taken to avoid any disturbance of the optical paths even, for example, by thermal air currents; in a laboratory situation this is not difficult to achieve and it has even proved possible to conduct c.w. laser holography in a factory environment by the use of careful techniques.

An alternative to such careful stability requirements is to reduce the time exposure to such a short period that the influence of all but the very highest frequency effects is eliminated. For this purpose a pulsed laser is used, capable of supplying adequate energy for the exposure of a photo-plate in one ultra-short pulse. This eliminates the need for laboratory conditions of stability. However, pulsed laser holography has certain drawbacks compared with c.w. work, namely (a) the limited coherence of pulsed lasers, (b) the difficulty of controlling the amount of energy delivered (and therefore, of controlling the photographic exposure), (c) the problems of alignment of the various optical components involved—usually overcome by the use of a small c.w. laser for pre-alignment and (d) they carry greater risk to personnel.

Generally speaking, holography can be pursued in almost any environment by the correct choice of equipment and techniques. A more adaptable holographic set-up than that shown previously is given in Fig. 4 for use with a c.w. laser. The use of a beam-splitter allows us to arrange the object and reference beam positions more suitably; thus, by having the reference beam well off-axis the zero order of the reference beam in reconstruction passes well clear of the eyes so providing a clearer view of the reconstructed image, the first order diffraction (see also Fig. A1). Also, it is easy to adjust the relative intensities of the two beams by the right choice of beam-splitter. If this is a variable beam-splitter then all kinds of permutations can be made. The usefulness of this is that the beam ratios are best arranged to suit the object being holographed. If the object is of very uniform reflectance the two beams can afford to have near equal strength as they arrive at the photo-plate. As the object surface's reflectivity becomes increasingly varied so it becomes necessary to increase the relative strength of the reference beam so as to prevent it being overmodulated by the object beam. For an average object a suitable beam ratio measured at the photo-plate might be about 5:1, reference to object. For an object with some highly reflective areas it would be desirable to increase this ratio even

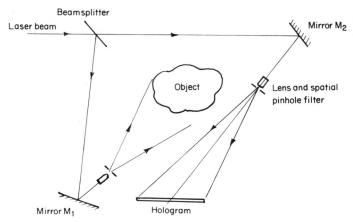

Fig. 4. A more sophisticated holographic arrangement for general reflection holography.

further. The principle is that no single object ray should be stronger than its reference ray, yet for optimum fringe quality they should be near equal in strength. As we cannot measure single rays we have to make the compromise judgements described.

The spatial pinhole filters shown in Fig. 4 are used to clean up the laser light. They are usually used in conjunction with the lenses employed to diverge the laser beam from a c.w. laser. These lenses are invariably microscopic objective lenses which first focus the narrow laser beam down to the focal point of the lens whereafter it continues as a divergent beam. A small pinhole in a mount is arranged at the focal point so that, as shown in Fig. 5, only the clean laser light passes through the pinhole while any other

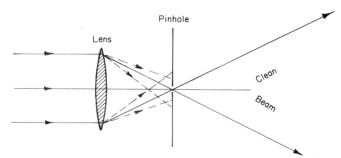

Fig. 5. A pinhole, carefully arranged to be at the focal point of the lens, which blocks off unwanted diffraction effects, rogue wavelengths, etc.

odd wavelengths, diffractions from artifacts, etc., try to focus elsewhere and are blocked by the mount. Such filters are not strictly necessary but they help produce much better holograms.

4. HOLOGRAPHIC INTERFEROMETRY

For most readers the principal interest in the application of holography will probably be in the field of holographic interferometry, and so this particular area will be examined first, and in some detail, while other uses are left until later; this makes more sense than might be thought at first as a grasp of the principles involved in holographic interferometry makes all other applications much easier to understand.

An examination of the mathematical description of the holographic process given at the end of this chapter shows that, apart from a possible change in brightness, the image diffracted out of a hologram is an exact restatement of the original set of light waves reflected by the object. Around 1964–65 Powell and Stetson[5] and others realised that this set of waves could be made to interfere with the actual set of waves which would again appear if the black card that was used to block off the object beam (p. 164) was removed. Then, if we do this we will see, effectively, two objects: the real one plus its exact image. Now suppose the object is given a small deformation. Its descriptive light rays will no longer match exactly those stored in the hologram and so interference between the two sets will take place. Making sense of the resulting interference pattern is often very difficult and sometimes even impossible but, even so, the judicious use of appropriate techniques can give useful answers to many problems.

Let us start with something simple. Suppose we have made a hologram of a cantilever and that it has been arranged so that the object beam, loading direction and observation direction are all normal to the cantilever as shown in Fig. 6. After the application of a small load to one end of the canti-

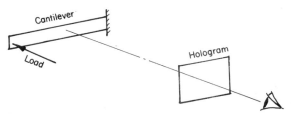

Fig. 6. Loading, illuminating and viewing a cantilever normal to its surface.

lever the observer, on looking through the hologram, will see two cantilevers superimposed on each other: the original, stored in the hologram, and the bent one. At the fixed end they will match exactly and so appear as one. As the eye moves along the cantilever away from the fixed end the two sets of light rays arriving at the eye will get further and further out of step until a point is reached where they are in exact antiphase. If the two images have been arranged to be of equal brightness at the start then this point of antiphase will appear as a dark fringe. If the distortion of the cantilever is not large then the dark fringe will be sufficiently broad to show up clearly to the naked eye as the rays on either side of exact antiphase approximate closely enough to perfect destructive interference. Continuing along the cantilever the rays will come back into step again giving a bright fringe, then another point of antiphase—that is, another dark fringe—and so on. So the cantilever will have a series of dark fringes along its length, each fringe being normal to the length of the beam. A photograph taken through such a hologram of a cantilever is shown in Fig. 7.

It is easy to see that each successive fringe represents a point at which the difference in the path lengths of the two sets of light rays has changed by one complete wavelength. As the object illumination has to go to the cantilever then return to the eye we see that the increase in path length of the rays describing the bent cantilever will be twice the displacement involved. If the light has a wavelength λ then the first fringe will occur when the path length of the rays describing the bent cantilever are $\lambda/2$ longer than those stored in

Fig. 7. Hologram of 8 in long cantilever fixed at its left hand end subjected to a point load at its right hand end. An inch steel rule is shown alongside. Note the slight angle of the fringes indicating that the load was not perfectly aligned on the mid-axis of the beam. The magnitude of deformation at any point is easily determined by counting fringes from the fixed end.

the hologram describing the unloaded beam; this will happen where the cantilever displacement is $\lambda/4$. The next fringe will occur where the deflection of the beam is $3\lambda/4$, the next at $5\lambda/4$, and so on.

Here, then, is an excellent system for measuring the cantilever deflection with great precision. It is not without its snags, however. For instance, successive fringes represent one wavelength difference in path length but they do not indicate (at least, directly) whether it is an increase or a decrease in path length. In other words, the fringes do not indicate directly whether the cantilever is bending towards or away from the observer or even going through an anticlastic point. Also, the fringes give a vectorial resultant of *all* displacements which occur between the before and after loading states; any rigid body movements, lateral displacements, etc., will also contribute to the production of each fringe in a complicated way.

5. FRINGE INTERPRETATION

The successful use of holographic interferometry clearly depends on an understanding of, and an ability to interpret, fringes. In examining the deformation of a cantilever the authors set up one of the simplest of problems to which holography might be applied. The resulting fringes were easily interpreted—particularly as they knew what to expect! In a more general situation it is not that simple. Consider the situation shown in Fig. 8

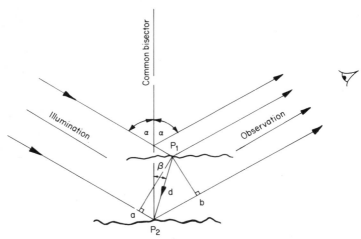

Fig. 8. The geometry of fringe formation due to displacement.

in which a point of an illuminated surface is displaced from a position P_1 through a distance d to a position P_2. Let the illumination and observation directions make an angle 2α with each other and let the displacement d make an angle β with their common bisector. If the total increase in the path length of the rays is Δ then:

$$\Delta = aP_2 + P_2b$$
$$= d[\cos(\alpha + \beta) + \cos(\alpha - \beta)]$$
$$= 2d\cos\alpha\cos\beta$$

That is, the system is only sensitive to the component of the displacement lying along the common bisector (the $\cos\beta$ component) of the observation and illumination directions. It has no sensitivity to displacements normal to the bisector. (Incidentally, this analysis shows that the maximum sensitivity of the set-up is achieved by making α as small as possible.) This is not the whole truth, however. In setting up the simple analysis just given no reference was made to the third dimension. Without defining it the implication was that the authors were considering the geometry of a displacement in the common plane of the illumination and observation vectors. While a good many problems either are, or can be arranged to be of this kind, the more general case will involve displacements in all three dimensions; moreover, in addition to simple translations a body may also suffer rotations. For a more general analysis of three-dimensional displacements the reader is referred to the various textbooks and papers in the reference list and particularly to the excellent paper by Burch[6] which is easy to follow and a splendid base for further study.

6.   TECHNIQUES

There are essentially three main techniques or methods used in holographic interferometry: single exposure, double (or multiple) exposure and continuous or time-averaged exposure. In the first of these, one condition or state of the object is recorded in a hologram and subsequently compared interferometrically with later states. This requires that the hologram be very precisely relocated after processing—a fairly difficult thing to do. An alternative is to process the hologram without moving it—*in situ*—but this also can be awkward. The advantage of this method is that various 'live' changes of state can be studied against the original.

Double exposure holography is much easier to use and has many advantages over the single exposure technique. It consists of making a hologram of one state of the object, changing to another state (by loading it, for example) and making a second holographic exposure on the same plate before the plate is processed. The two exposures will usually be exactly equal, photographically, and so the interference fringes formed between the two states will have optimum contrast. As there is no relocation problem the only fringes formed are those relating to the change of state of the object. Once recorded, the fringes can be studied—anywhere—without fear of losing them. The only drawback is that only one *past* history can be examined, although several variations exist which allow for more than one change of state to be recorded.

Time-averaged holography is used to examine regular motion, particularly vibrations. This is one of the most easily and profitably used methods in the study of vibration and was one of the earliest applications of holographic interferometry, being first developed by Powell and Stetson[5] and others around 1965. If a hologram is made of an object while it is vibrating over a period of time sufficient to include many cycles of the vibration it will show fringes which are contours of equal amplitude of vibration and bright areas which are nodal areas (see Fig. 9). The hologram may be considered as being like a double (or multiple) exposure hologram in which each alternative exposure is of one of the two extreme positions of the vibration cycle—rather like a time exposure photograph of a swinging pendulum which tends to show only the two extreme rest positions of the pendulum, the in-between positions blurring out. This is not an exact

Fig. 9. The same cantilever as in Fig. 7 but this time in resonant vibration. A vibration transducer is fixed to the same loading point as before and driven by an oscillator through an amplifier. This time-averaged hologram shows the central node corresponding to the fundamental torsional frequency of the beam. The amplitude of vibration at any point is easily determined by counting fringes from the node out to that point.

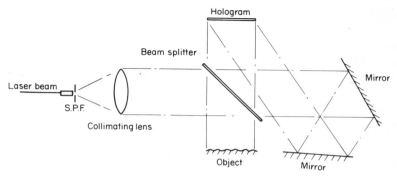

Fig. 10. A holographic set up which may be used to achieve illumination and viewing normal to the surface of a plane object. S.P.F. = spatial filter.

analogy for what happens but it helps! Whatever technique is employed either of the two set-ups previously described may be used, that shown in Fig. 4 being the more adaptable and being commonly used, especially for pictorial or qualitative|holography. When maximum sensitivity to out-of-plane displacements is required the set-ups shown in Figs 10 and 11 may be used. The first of these is suitable for cases when the object is not exceptionally large and provides for normal illumination and viewing of the object, thereby giving maximum sensitivity to normal displacements. At

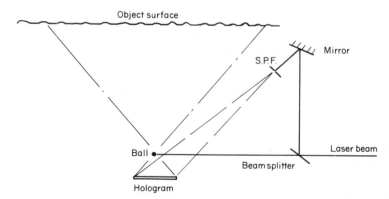

Fig. 11. An economical method of making holograms of a large surface area. Two economies are achieved: first, the use of a small shiny ball is a cheap means of spreading the laser light over the object surface; second, the use of retroreflective paint on the object surface ensures that almost all of the light that illuminates it is used in making the hologram so allowing a low power laser to be used. S.P.F. = spatial filter.

first sight it may look as if there is one mirror too many in this set-up; this is to allow for the plane-polarisation that exists in the output of many lasers. When using plane-polarised light it must be remembered that the plane of polarisation is rotated through 90 ° on each spectral reflection; thus, the numbers of reflections of the object and reference beams should both be either even or odd together for optimum interference to be displayed. Of course, if the object has a very diffusely reflecting surface or if the output of the laser is not plane-polarised then this does not matter and one mirror less may be used.

If a very large object is to be examined the system shown in Fig. 11 may be used. Here, the object illumination is derived by reflecting the laser output beam directly off the spherical surface of a ball-bearing! If the surface of the object is coated with retroreflective paint then almost all of the surface illumination is reflected back towards the ball; as the retroreflection is not all that precise the reflected light will by-pass the ball and land on the photo-plate, giving a high degree of efficiency in the use of the light available. The presence of the ball in the reconstruction is a trivial nuisance, being easily overlooked.

There are many other set-ups that may be described but space precludes this. It is easy to invent one that suits the particular application in any given case (for example, substituting a roller bearing for a ball bearing in the above case gives a long narrow beam for the object illumination!) and, in any case, the literature is readily available describing many specific situations.

7. SELECTION OF EQUIPMENT AND PRACTICAL SUGGESTIONS

7.1. Continuous Wave Holography

The choice of a laser depends on the type of work to be done, how big an area is to be covered and how deep a volume is to be examined; also, the stability of the environment in which the work will be done has to be taken into account. If it can be guaranteed that there will be no movement of any kind whatsoever in the working locality, that it will be entirely vibration free, then very large areas can be covered with a very small laser and long time exposures. In most cases, however, there are usually enough vibration problems to encourage the use of at least a few milliwatts of laser power. A lot can be done with about 3 mW, life is easier with 15 mW and very comfortable indeed with 50 mW. Higher powers than these may be desirable in some cases, particularly where short exposure times are required; for example, stroboscopic vibration analysis often employs lasers of several watts.

Remembering that the coherence length of the laser dictates the maximum path difference that can be allowed between the object and the reference beams, it is obvious that this also defines the maximum depth that is permissible in the scene to be holographed. A typical value of coherence length of a helium–neon laser might be 25 cm but some lasers offer less and some very much more (> 10 m, say).

Most holographers like to work on a large heavy table mounted on some kind of anti-vibration support. It is not necessary to have a very smooth surface (in spite of what the glossy advertisements say!) but a large mass on soft sprung cushioning tends to eliminate the effects of extraneous vibrations. In the authors' laboratory an old steel surface table bought from a local scrap dealer is used. It is about 6000 kg and sits on six air-cushioned legs. The advantage of the steel top is that all the optics can be held firmly in place, once set, on magnetic stands. The laser can afford to be separated on another bench which does not even need to be exceptionally stable (though it should not be too unstable either!). From the first beam splitter onwards stability is essential.

Having put this in print will no doubt bring forth comment from others claiming to perform successful holography on the kitchen table with a one candle-power laser; that seems neither hygienic nor sensible. Stability removes doubt, if nothing else.

For most applications it is not necessary to spend a fortune on mirrors and beam-splitters. On the contrary, ordinary float glass bought from a local glazier can be coated to give results which are often superior to the optically worked ones which are so expensive. A large diameter beam often has an irritating orange-peel effect when reflected by an optically worked surface. If the local glazier is asked to select a piece of float glass away from the edges of a main sheet and as scratch free as possible this will be perfectly adequate. One objection to this method of obtaining cheap mirrors and beam-splitters is that the coating tends to be soft and difficult to clean. However, with reasonable care they are very successful and, in any case, at this sort of cost they can be considered disposable.

The lenses used for diverging the laser beam (from a c.w. laser) are almost invariably microscope objective lenses, popular sizes being 10 ×, 20 × and 40 ×. As has already been explained, it is common to insert a pin-hole at the focus point of the lens so as to obtain a clean beam. Arranging the lens and pin-hole together in the laser beam can be difficult. For a 20 × lens, for example, the optimum pin-hole will be 10 μm and this has to be adjusted until it centres exactly at the focal point of the lens. It is probably best to obtain ready-made units for this purpose and these are usually sold as combination units on stands and are called spatial pinhole filters (although this term properly refers only to one part of the unit).

The next item of equipment required is a hologram plate holder; this is readily made in most workshops. For pictorial holography, or for double-exposure and time-averaged holography it matters little how the photographic plate is held as long as it is supported rigidly during the exposure. However, for live-fringe viewing it is usually necessary to take the plate away for processing and then to return it, relocating it precisely in its holder. For this purpose a kinematic mount may be used, the plate being supported on its lower edge by two pins and on one vertical edge by one pin; lateral location can be provided by three small ball bearings against which the plate is *gently* pressed. Abramson[7] suggests allowing the plate to have a small tilt backwards so that gravity holds it in place against the various supports; this avoids any problems from mechanical stressing of the plate.

Mirrors, beam-splitters, spatial pinhole filters, etc., must not move during exposure for any holography and, in the case of live fringe work they must not be moved subsequently either. For this reason a steel surface table makes a splendid worktop for holography as all of these components can be supported on magnetic bases which clamp firmly to the table. The usual magnetic toolroom base is ideal for this purpose and can be purchased separately.

Other odds and ends that are useful to the holographer include any kinds of lenses, particularly large diameter ones. Old aircraft cameras are a useful source of these. Also magnets for holding things in place, lab jacks for adjusting object heights, etc., and an old camera shutter for timing exposures. As these are likely to be measured in seconds from time to time the shutter should have a B or T setting.

The last piece of equipment which might be considered necessary is an exposure meter. In some cases it might be worth investing in a laser power meter (usually sold by the laser manufacturer) and this may be suitable for use as an exposure meter too; but again power meters tend to be very expensive things and are usually awkward for measuring exposures while only being rarely used in adjusting the laser. For measuring holographic exposures an ordinary, good quality photographic exposure meter is perfectly adequate. The meter used by the authors is the Lunasix. With the incident light dome in position the meter is placed at the location of the

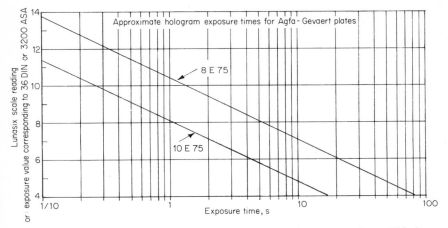

Fig. 12. Calibration chart used to estimate holographic exposure times. This one may be used for a first attempt, but it has aged somewhat now and the plates referred to may have been upgraded in speed.

photo-plate for a few seconds and the needle locked, the reading being then taken with the laboratory lights on. Figure 12 shows the calibration chart used by the authors for Agfa-Gevaert holographic plates. Obviously, this chart is used with a discretion that is born of long experience but it should, at least, give a fair starting point. (Needless to say, the process is not as log-linear as the straight lines suggest; this is only a fair approximation.)

7.2. Pulsed Laser Holography

The choice of a pulsed laser is more complicated and, in a sense, more restrictive. The output of a single pulse laser is usually a discrete quantity of energy with only a limited range of control over the quantity; in other words, for most purposes it is usually not possible to adjust the time of exposure in order to get an adequately exposed plate. This effectively restricts each pulsed laser to a limited range of volumes it can cover. Other factors affect this too, of course, such as the speed of the photographic emulsion and the reflectivity of the object. However, it is wise to seek direct help from an established user in selecting a pulsed laser for a particular area of work. (That is probably good advice for those selecting a c.w. laser too.)

For pulsed laser holography environmental stability is hardly ever a problem. As the pulse length is so short—perhaps a few nanoseconds—vibrations, bumps, etc., are of less significance. Even so, it is helpful to prevent the accidental movement of apparatus once set in position and, again, a steel table with the optical equipment mounted on magnetic bases is useful. Vibration isolation may not be necessary in this case but it is available when the worktop is also used for c.w. holography.

Getting a divergent beam from a pulsed laser introduces a new problem. If the beam from the laser is converged first, as in the case of the c.w. laser, then where it passes through the focal point of the lens there is momentarily an enormous amount of energy. This is likely to greatly disturb the optical properties of the air at this point and would certainly blast the centre out of any pinhole placed there! So, instead of converging with positive lenses it is necessary to diverge immediately with negative lenses. Of course, this means that a spatial pinhole filter cannot be used and so the beam cannot be cleaned up. A good source of cheap, suitable lenses is an optician's trial kit, which can often be purchased in perfect condition secondhand at great savings.

An ordinary photographic meter has no value for exposure measurement in pulsed laser work; instead, exposure tends to be more a matter of trial and error. It is possible to calculate the exposure requirements and this will probably be done to determine the probable laser size required but thereafter a few tests will probably be quicker and more satisfactory.

7.3. Recording Materials

The reader must be referred elsewhere to read about newer methods which use thermoplastics, photochromics, dichromated gelatins, etc. These all offer exciting prospects. However, the large bulk of holography is still carried out using silver halide photographic materials and will probably continue to be so for a long time to come. A newcomer to the business would be well advised to stay with these at least until he knows what he is about. An excellent description of the various alternative holographic recording media is given by Smith.[8]

7.4. The Laser Laboratory

The laboratory should be reasonably light tight and is best ventilated by convection rather than a fan as strong air currents can upset the more delicate holographic work. A useful idea is to have string running just overhead around the table and connected to a pull/pull light switch. This allows the lights to be controlled from anywhere in the laboratory.

Holographic plates are relatively slow and so it matters little what colour the laboratory walls are. A light pastel shade of green or grey is useful in allowing a little visibility to the user in the reflected light of the laser.

It is important for safety that the laboratory should not have any shiny objects where the laser light may land. This is imperative when working with a pulsed laser. Even a low powered laser can cause permanent eye damage if its undiverged beam is allowed to enter the pupil of the eye and carelessness could result in blindness. When working with a pulsed laser there is an even greater risk and every precaution must be taken to safeguard anyone who might enter the laboratory. When making straightforward holograms with a pulsed laser there is nothing to watch anyway, so it is best to completely mask the eyes. If there is a need to look at something during the exposure then this should be done in a screened-off area.

Illuminated signs connected with the laser may be placed out of doors as a warning that a laser is in use and it should be noted that the safety regulations being introduced in many countries require warning signs for laser radiation. Also, locks which keep people out while a hologram is being made are almost essential if exposures are not to be ruined, but such locks should have an external override provided for safety.

Having given these warnings it must be said that it is easy to avoid accidents by the use of sensible care and one should not be afraid of using a laser.

8. AₚPLICATIONS OF HOLOGRAPHY

The variety of uses for holography is so immense that it is impossible to list all of them here, even in index form. Indeed, it is true to say that new applications appear just as fast as the imagination can work! The aim in this section is to present a selection of important applications of holography to illustrate the scope of this very versatile technique. The examples chosen are almost a random selection from the very many problems to which holography has been successfully applied as space precludes a wider coverage and for the same reason, detailed descriptions of the various examples have been omitted.

8.1. Holography as a 3-D Imaging Technique

8.1.1. Displays
The three-dimensional property of the hologram has been described above and although as an optical transducer it is the extension of the fundamental

technique to holographic interferometry that is of primary importance, the use of holograms as 3-D photographs has a role to play in some industrial situations. Because of its ability to faithfully and precisely reconstruct the three-dimensional image of an object, holography has found a place in advertising. Recently an American company sent a representative to an exhibition in the UK with a range of their products in the representative's brief case in the form of a dozen or so glass plates. Some of these products could hardly have been carried by one man let alone placed in a brief-case, so one benefit of a hologram is obvious: upon subsequent reconstruction at the exhibition, the products could be rigorously inspected by eye—but not touched.

8.1.2. Volume photographs

Engineers at the CEGB Research Laboratories, Southampton, are using the three-dimensional properties of holograms to record such items as fuel elements from the Advanced Gas Cooled Reactor to facilitate improved optical inspection.[9]

By making a hologram of the fuel element looking along its axis, it is possible to make a very detailed examination of any section along its length from the subsequent reconstruction of the real image. Figure 13(a) shows the end caps of the elements and the front bracing, which is out-of-focus in Fig. 13(a), is clearly displayed in Fig. 13(b). The location of any other plane within the image is easily achieved and Fig. 13(c) shows the effect of focusing on the centre bracing. The amount of detail stored in the holographic recording can be judged from the definition of one small part of the front bracing (Fig. 13(d)).

This volume aspect of the holographic image has proved to be of use in the study of such subjects as tracing atomic tracks in bubble chambers[10] and particle sizing of the droplets produced in aerosols,[11] amongst others. In the former case the hologram has the distinct advantage over conventional imaging systems where these have a relatively short depth of field and hence important information might well be lost by a poor choice of the plane of focus within the chamber. Reconstructing the holographic image permits the entire depth of the chamber to be scanned in a similar manner to the reactor fuel element in Fig. 13.

In the analysis of aerosol particles one major problem is again the limited depth of field, introduced in this case by the high magnification optics necessary to resolve the fine detail of the particles which may be of the order of a few tens of microns in diameter. Thomson[11] describes a simple in-line holographic arrangement which is used in conjunction with magnifying

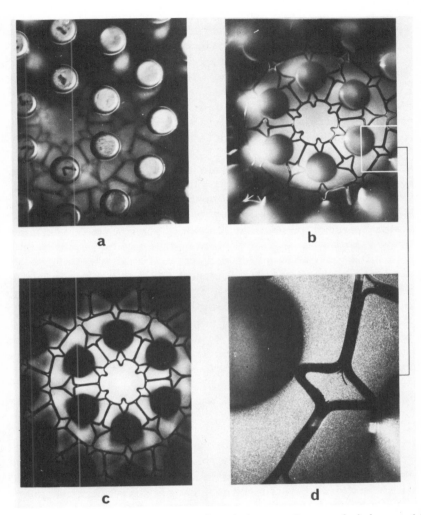

Fig. 13. Real image reconstructions from hologram of reactor fuel element: (a) end caps; (b) front brace; (c) centre brace; (d) detail from front brace. (Courtesy of Marchwood Engineering Laboratories, CEGB Marchwood, Southampton.)

lenses and a television camera to produce a × 300 magnified image of the sample spray volume on a television monitor. The image is amenable to inspection throughout its depth and by using a sufficiently large aperture in the viewing optics, discrete planes can be selected giving a high resolution of the individual particles. In both these examples, Q-switched pulsed lasers would be required to construct the holograms because of the short duration of the phenomena.

More recently Briones[12] has described a holographic system for examining water droplets ejected from a nozzle at sonic velocities whereby a scene depth of some 380 mm could be utilised with a droplet resolution limit of a little over 3 μm.

The study of aerosols can be extended further by using double-exposure techniques, that is, recording two images of the particles on one holographic plate, with an appropriate time separation between the exposures. This yields information on the velocity and direction of individual particles and is easily accomplished using the present generation of pulsed lasers which are capable of emitting single, double and multiple pulses with known, and controllable, pulse separation. This principle has been applied to the measurement of fluid flow and has become widely recognised (see, for example, ref. 51). Small particles can be injected into the fluid stream and double (or multiple) holograms are recorded with a time interval of tens or hundreds of microseconds between the exposures. The composite image of the fluid stream can then be examined and the distance moved by any particular particle measured, again throughout the entire volume, to determine the stream velocity in that region.

8.2. Some Applications in Physical Science

8.2.1. Microscopy

One application of holography which may benefit the physicist has been given above, viz. the work on bubble chambers, but the purpose for which holography was originally invented—microscopy[1]—has met with limited success. Nevertheless, work is still continuing in this field and holography has an important role to play because of its volume characteristics, already discussed. Much of the pioneering work was carried out by van Ligten and Osterberg[13] and a good summary of the approaches can be found in the text by Collier *et al.*[14] Although the resolution seems comparable with conventional optical microscopes, holography opens up the possibility of microscopically examining transient behaviour and, of course, in a three-dimensional way, the work on particle analysis falls within this category.

The extension of the technique to the Angstrom resolution range still seems quite far off.

8.2.2. Plasma studies

Electron density measurement in a theta-pinched plasma using holographic interferometry has been described by Nicholson *et al.*[15] In this study, the electron density is integrated over the length of the plasma by holographically recording an axial view along the discharge tube. The information in such holographic reconstructions is usually poor because of the integrating effects from different directions along the tube but this was overcome by selecting unidirectional light from the image, that is, using a telecentric viewing system. Figure 14 shows reconstructions from a hologram of the plasma; Fig. 14(a) is a real image reconstruction using all of the light field; Fig. 14(b) shows the fringes formed in the telecentric system where the light field is axial with respect to the discharge tube; Fig. 14(c) shows the effect of reconstructing at an angle of 0·012 radians with respect to the axis. The interference fringes represent a change in optical path length and this in turn is related to the electron density. The error in the measurement of the density was quoted as less than 2 %.

8.2.3. Interferometry

Having just discussed the use of holographic interferometry as a tool for the plasma physicist, it is worth noting that holography has opened up new possibilities in the field of interferometry itself. In making a comparison between classical and holographic interferometry it could be true to say that the former is, in fact, only one specialised case of the latter. Holography has extended the scope of classical interferometry by making it possible, for example, to create interference effects on diffuse surfaces and to greatly simplify the procedures involved in the classical arrangements because there is less critical alignment involved.

8.2.4. Lens testing

Holography has also come to the assistance of the lens maker by making the inspection of lenses, at various stages of their fabrication, a much less costly exercise. Conventionally, the lens being worked is compared against a standard, precision test glass and interference effects (Newton's rings) are observed between the closely matched surfaces. This requires a number of test glasses to be available for each lens, hence the high cost involved. By making a single holographic recording of a master surface, subsequent lenses can readily by inspected by optically superimposing them on the

E. R. Robertson and W. King

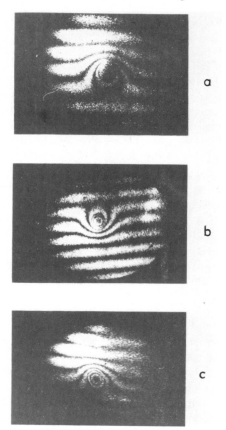

Fig. 14. Theta-pinch plasma studied by holographic interferometry: (a) real image reconstruction; (b) axial light field; (c) off-axis light field. (Courtesy of J. P. Nicholson, University of Strathclyde.)

reconstructed holographic image of the perfect surface. This idea has been taken further by Hildebrand[16] who has proposed a method in which the lens surface can be worked to the desired form without the necessity to remove it from the polishing tool, which was one of the drawbacks with the above interferometric comparison against the master surface.

Further examples of holography in optical physics, such as its use for correcting aberrations in lenses and as image filters, are outside the scope of this text but the reader is referred to any of the more specialised texts on holography (for example, ref. 14).

8.3. Some Examples in Engineering

Almost all applications of holography in engineering are based on the principles of holographic interferometry and from this, displacement and strain measurement, determination of shapes, non-destructive inspection and a host of other problems have been treated very successfully. The fundamental idea of holographic interferometry occurred to a number of researchers at the same time (e.g. refs 5, 17, 18) and since then the field has expanded at a staggering rate. The following examples are chosen from areas in which holography is already well established and where the potential for the technique looks hopeful.

8.3.1. Strain and displacement measurements

As explained previously, holographic interference fringes represent the displacement of a point along the bisector between the illumination and observation directions. Change these directions, e.g. by moving the eye behind a hologram, and the bisector changes, causing a change in the displacement component and resulting in a change in the fringe pattern. In general, one must obtain fringe patterns for three separate and non-planar (the three must not lie in the same plane) bisectors in order to obtain three independent components of displacement in space. In practice, one is often interested in either the out-of-plane displacements (measured along the normal to the surface) or the in-plane displacements measured perpendicular to the normal, particularly for flat surfaces.

For out-of-plane displacements alone it is only necessary to arrange the illumination and observe directions to be equally and oppositely inclined to the surface normal, with maximum sensitivity being obtained with the arrangement of Fig. 10. To obtain in-plane displacements only, as required for strain measurement applications, a more complex arrangement is necessary.

Ennos[19] was the first to attempt experimental separation of the in-plane and out-of-plane displacements. He used two holograms inclined at equal and opposite angles to the surface normal, with the surface illuminated by a single beam. By vectorially subtracting the displacements measured from these two holograms, he cancelled the out-of-plane components and was left with the in-plane components (it should be appreciated by the reader that it is almost impossible to arrange the mechanical testing of a structural component without introducing quite large out-of-plane displacements). This basic arrangement was simplified subsequently by Butters,[20] Luxmoore and House[21] and Boone,[22] who all used a single hologram placed normal to the surface, but with two illuminating beams inclined at equal but opposite angles to the normal. This arrangement produced a dual fringe pattern on the hologram, representing both in-plane and out-of-plane displacements. The presence of two fringe patterns complicated the analysis, and further developments utilised the speckle effect (see Chapter 7). For example if one uses a lens to focus the object onto the hologram while making the hologram (referred to as 'focused image holography') then both in-plane and out-of-plane displacement patterns can be separated by suitable optical filtering of the doubly exposed hologram.[23,24]

The use of focused image holography simplifies two of the major problems affecting the practical use of holography for quasistatic measurements, namely, vibration and rigid body movements. Instead of using a lens to focus the object onto the hologram, the holographic plate is placed on the object, with emulsion side down, and illuminated from behind. The direct beam passing through is the reference beam and the light scattered back by the object is the signal beam. This technique has been referred to as reflection holography,[23,24] and the closeness of hologram and object reproduces surface detail of the object just as a lens would. Mounting the plate on the object removes most of the problems associated with vibration and rigid body movements, and holograms can be made in industrial environments.

If the incident laser light is normal to the hologram plate, then the fringes are sensitive only to out-of-plane displacements, and Ennos and Virdee[25,26] have used this technique in several practical problems to obtain holographic interference fringes. In-plane displacements can be computed by illuminating from several oblique directions, but they suggest the speckle technique is more advantageous for direct in-plane measurements. A detailed theoretical comparison of holographic and speckle methods for strain measurement is given by Dandliker.[27]

8.3.2. Vibration studies

It is probably true to say that holographic interferometry brought about one of the most significant advances in the experimental study of vibrations for many years. In fact, this particular application can be thought of as having been as beneficial to holography because it gave a great deal of credence to the technique which had been thought of simply as an attractive toy for scientists to play with. Its popularity in this field is largely attributable to the relative simplicity of the technique and the way in which it can readily describe the vibrational characteristics of any surface in both a qualitative and quantitative way.

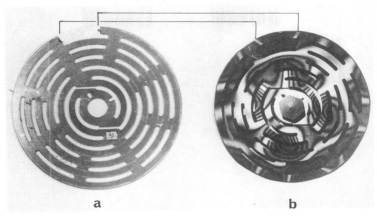

Fig. 15. Vibration study of a compressor valve cover plate: (a) a fractured plate; (b) time-averaged vibration pattern, 2·7 kHz. (Experimental work by R. Weir.)

Perhaps the simplest of the holographic interferometric techniques to apply to vibrating surfaces is the time-averaged method (described above). An example of this is shown in Fig. 15 from a study on compressor valve cover plates.[28] As the valve opens and closes the cover plate is subject to a severe vibration which ultimately leads to fatigue failure (Fig. 15(a)). By subjecting a cover plate to a range of frequencies it is possible to rapidly establish the resonances using real-time, time-averaged holography and so predict possible failure conditions. Figure 15(b) shows one resonant condition which could well account for the fracture of the plate in Fig. 15(a)—note the position of the nodal (bright) regions. Figure 15 also serves to illustrate one of the disadvantages of the time-averaged technique, that is, the loss of contrast in the interferometric fringes with increasing vibration amplitude. From a purely qualitative point of view this is of little consequence, because the nodal regions will always be readily identifiable, but it makes a quantitative assessment of the vibration amplitude very difficult if not impossible.

The reduction in fringe contrast with increasing amplitude is a direct result of the time-average recording and it can be shown that the fringe intensity, I, is of the form

$$I \propto J_0^2(\alpha)$$

where J_0 is the zero order Bessel function and α is a function of the displacement amplitude and the geometry of the holographic arrangement (see, for example, ref. 5). From this relationship it can be shown that with a

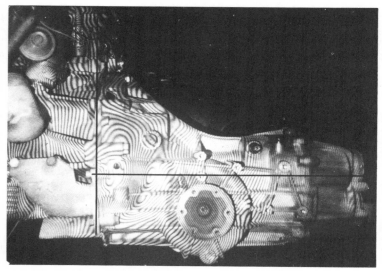

Fig. 16. Vibration pattern of gearbox at 6200 rpm—double exposure using a pulsed laser. (Courtesy of Felske and Happe, Volkswagenwerk AG, Wolfsburg.)

fringe order number of 5 (equivalent to approximately 1 μm displacement amplitude), the fringe intensity is reduced to around 5 % of the peak value.

To overcome this problem two approaches can be made, (i) using the double exposure method in conjunction with a pulsed laser, or (ii) stroboscopically freezing the vibration pattern and using the double exposure method. Of the two, the former is more widely used partly because of the increase in availability of double pulsing pulsed lasers with high coherence characteristics and also because the optical system is usually simpler. The main disadvantage in gaining fringe contrast is that the nodal lines are no longer clearly visible. This points to the complementary nature of the two approaches.

Felske and Happe[29,30] at Volkswagenwerk AG (West Germany) have used a specially designed double pulsed hologram camera* to study a number of problems in automobile components which are subject to engine vibrations. Figure 16 is one example of this work, on the measurement of stiffness of a gearbox.[29] From their interpretation of the vibration fringes

---

* A hologram camera, or holocamera, is simply a laser and the appropriate optical components and plateholder mounted on a suitable surface which is then capable of being transported from one place to another.

they concluded that the stiffness in the region of the clutch housing was inadequate and this led directly to an improvement of the design. Further work by the same authors on automobile disc brakes has led to the detection of vibration sources which were responsible for brake squealing.[30]

The use of hologram cameras has greatly improved the potential applications of holography because they allow the technique to be taken to the problem rather than having to take the problem to the holography laboratory. Developments in this area have even reached the stage where holograms can be constructed outwith the confines of the darkened room and even in the presence of high ambient lighting. This was demonstrated by Erf *et al.*[31] who conducted tests on a helicopter in a flight hanger, again using a double pulsing laser. To stop fogging of the photographic emulsion under these conditions a shutter is used which is opened for approximately 75 ms which is more than sufficient time for the two laser pulses to enter and expose the plate, the pulse separation being at most 1 ms. Figure 17 shows the vibratory effect on one section of the air frame due to an impulse excitation from a rivet gun. The buckling characteristics of the airframe within the unsupported regions are clearly visible.

In many cases of vibrating subjects, the subject is not in a fixed position when it is being excited. There are occasions where the body is rotating and this often leads to resonances being set up. An example of this is in engine or turbine blading systems. The rotational effect of such systems can often be optically eliminated by employing what is known as an image-derotator[32] and is effective for rotational speeds of up to 35 000 rpm.

8.3.3. Inspection methods

The examination of structures using holography as a qualitative inspection technique has been gaining in popularity for a number of years and offers a unique solution to many problems.

In laminated structures, for example, it is often very difficult to look inside the body and detect the presence of disbonding between layers. One excellent example of this is in the inspection of pneumatic tyres where there is a multiplicity of plies within the structure held together by rubber cements. A failure in the bonding of the plies could lead to disastrous consequences. By using double exposure holographic interferometry the debonds manifest themselves as regions of high fringe density in an otherwise low fringe field. The hologram is double exposed using a different internal pressure in the tyre for each exposure. The debonds are highlighted because of the reduced stiffness in these areas.

Dynamic loading can also be used as a means of detecting flaws in

Fig. 17. Reconstruction from a holographic interferogram of a section of a helicopter air frame following dynamic excitation from a rivet gun. (Courtesy of R. K. Erf, United Technologies.)

structures. Any surface discontinuity will cause changes to take place in the fringe frequency in, say, a uniform fringe field or cause fringes to be present in an otherwise fringe-free area. Holography has opened up a new possibility for this type of non-destructive testing where a specialised property of the double-exposed hologram (or interferogram) can be utilised to good effect—the fact that the interference fringes move across the image as the observer moves his viewing position across the hologram. Figure 18 is a sequence of views from a single interferogram showing a stress wave traversing a section of a rectangular bar containing a circular hole.[33] Any irregularity in the movement of the fringe pattern as the observer scans the hologram will indicate the presence of some

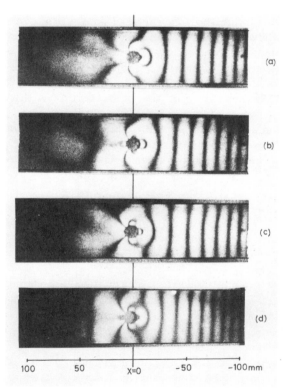

Fig. 18. Observation of a transient stress wave (moving right to left) in a rectangular bar with a circular discontinuity. The movement of the interference fringes by shifting the point of observation from right to left across the hologram ((a) to (d)) corresponds closely to the stress wave motion.

discontinuity on or near the surface. This system is currently under development to establish the magnitudes of defects which can be detected in this way.

Many specialised inspection techniques have been developed to study, for example, cylindrical bores, turbine blades, integrity of reinforcing spars (stringers) on aircraft wings and brazed, honeycomb-type laminated structures to name but a few. More detailed descriptions of these methods are given in refs 34–36.

8.3.4. Contouring

Surface topography is another area in which holography has much to offer, its principal advantages being the ability to work in small areas (small viewing apertures) and the wide range of contour intervals that can be generated. Two basic holographic methods are used to this end, (i) multiple refractive index (MRI) and (ii) multiple frequency (MF).

The former involves immersing the subject in a liquid or gas and making two holograms where the refractive index of the intervening medium has been changed between the exposures. In the second case, the laser used to construct the hologram (or contourgram as Robertson and Elliot[37] have called it), records one image at one frequency and the second at another frequency. The two methods are in fact very similar as can be seen from the expressions for the contour interval, Δ. In MRI the equation (for illumination and observation normal to the surface) is

$$\Delta = \lambda/2|n_1 - n_2|$$

where n_1 and n_2 are the refractive indices for the first and second exposure and λ is the wavelength of the laser source. For MF the contour interval is

$$\Delta = \lambda_1\lambda_2/2|\lambda_1 - \lambda_2|$$

but since $n_1 = \lambda/\lambda_1$ and $n_2 = \lambda/\lambda_2$ both equations are identical.

An example of MF holographic contouring is shown in Fig. 19. The contourgram is of a 50p piece and was constructed using two green lines from a krypton ion laser ($\lambda_1 = 520\cdot8$ nm and $\lambda_2 = 531\cdot8$ nm). The contour interval is $13\cdot8\ \mu$m.[37]

A novel method of holographic contouring was presented by Abramson using his sandwich holography technique.[38] A simple method of producing contours on a surface is to project a system of parallel fringes across the object surface, the projection being in the plane of the surface (e.g. Lloyd's mirror principle). The fundamental problem with this however is that high corrugations on the surface lead to large shadows which can mask much of

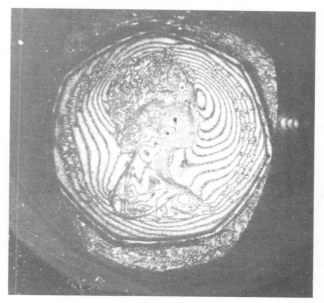

Fig. 19. Contour fringes on a 50p piece produced by multiple frequency holographic contouring; contour interval 13·8 μm.

the detail. Using the sandwich method the surface can be illuminated at a much steeper angle to its plane and by manipulating the two hologram plates forming the sandwich, the intersection of the contour planes with the surface can be altered to produce topographical maps whose axes lie along any direction. Contouring methods can also be used as a desensitising procedure for holography in the analysis of displacements because in some cases the holographic process is too sensitive to the object motion. By producing contour maps of the object surface before and after deformation, these can beat together to form moiré fringes showing the contours of displacement.[37]

Details of the use of contour maps for deformation measurement and a full description of the general contouring techniques can be found in ref. 3.

8.3.5. Fracture mechanics

Holographic interferometry has been used to detect the presence of cracks in a body in a qualitative way[34] and in a quantitative way to study fracture mechanics.[39] Static and dynamic double-exposure and real-time holo-graphic methods have all been applied to this class of problem and each problem has to be examined on its merits to decide which particular method should be adopted.

Figure 20 illustrates how holographic interferometry can clearly expose the presence of a surface crack in a beam which has been subjected to bending. The loading is such as to cause opening of the crack and the interference fringes are a result of the out-of-plane motion of the specimen.

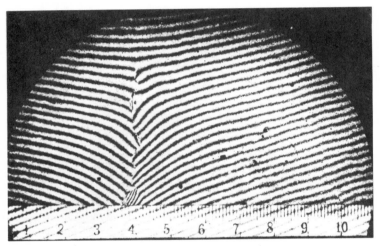

Fig. 20. Crack detection using real time holographic interferometry. (Courtesy of Pflug and Jacquot, EPFL, Lausanne.)

The system used here was a real-time one and it has been concluded[40] that this is the fastest and most efficient method for the detection of cracks but for a fuller quantitative analysis the holographic approach should be extended to real-time holographic-moiré.[41] Real-time viewing also has the advantage of being able to study the growth of the crack and Vest[42] has used this to monitor the effects of stress-corrosion cracking.

Cadoret[43] has demonstrated a real-time system for quantitatively studying the yield phenomena at a crack tip and the advantages claimed for the technique are that it is non-contacting, it can be used on any kind of material and surface (no need for preparation) and the measurements can be carried out simultaneously with the load effect. The plastic zone can be detected by the delineation of the elastic displacement fringes observed in the holographic image.

8.3.6. Thermodynamic studies

In thermodynamics, the measurement of density changes in gases and fluids is of great significance, examples occurring in heat transfer, supersonic flow, two-phase flow, etc. The use of holography to measure such changes has been examined by several workers including Kapur and Macleod[44] and Fraser and Kinloch.[45] Changes in density, however they are caused, result in a change in refractive index and this can be detected by the holographic process (because of the changes in the optical path length of the light forming the image). A major difficulty in quantifying the refractive index changes in a three-dimensional volume, particularly where these are asymmetric, is the inability even in a single hologram to look around the phase object—the observed interference fringes show the refractive index changes integrated along one particular direction. In order to build up sufficient information to be able to calculate the density changes a wide viewing angle is called for. Fraser[45] succeeded in doing this by recording the information on four hologram plates simultaneously, these spread around the object field and giving a 180° field of view. Kapur and Macleod[44] have described how holography can be used to measure the convective transfer properties of a heat exchanger which has a grooved surface—the type of surface preferred in heat transfer equipment. They claim that the method is unique in terms of the quality of the data that can be derived, both locally and overall, and it can be obtained very quickly and easily by comparison with other techniques. Work of this sort will undoubtedly be of benefit in cases such as nuclear reactor systems where localised reduction in heat transfer efficiency could lead to disastrous consequences. Guerry et al.[46] have successfully employed a real-time holographic system to study the convection transfer coefficients on the surfaces of solar panels and the authors state that the holographic approach was the most economical as well as the most easily applied approach to the problem.

8.4. Some Biomedical Applications

Many areas in the biomedical sciences have been examined using holographic techniques, these in a sense being extrapolations of many of the engineering applications discussed previously to particular biomedical problems. This is a rapidly expanding field and the interested reader should consult refs. 47–50 for more detailed information.

8.4.1. Ophthalmology

Direct holographic imaging of the eye is regarded as being a major step forward in fundus photography because of the greater depth of field and high resolutional capability over existing instruments. Ohzu has described the development of a fundus camera which is suited to both the conventional approach and the holographic one (p. 82, ref. 47). He concluded that it was easier to record a hologram with the apparatus than to take a normal photograph because the former does not require focusing.

Holographic interferometry also shows promise in being able to measure changes within the eye such as development of cataracts and retinal edema.

Contouring techniques in this area are also seen as having potential significance. Using MF holographic contouring it would be a relatively simple task to measure the front corneal surface or sclera for assisting the fitting of contact lenses, for example. The sandwich technique would also permit contouring in various directions and would greatly assist in the measurements for this type of application. Greguss[50] also suggests the contouring technique might be used as an early warning of the threat of glaucoma.

8.4.2. Odontic applications

Holographic interferometry has been employed in this area to study the behaviour of both natural teeth and prosthodontic appliances. The mobility of teeth can be measured by this method *in vivo* and also the thermal behaviour of enamel, denture and filling materials. This latter example being of importance in establishing differential expansions which could take place between the tooth and the filling leading to separation of the two.

Contouring of gums is also a proposed study to establish the shrinkage which takes place following teeth extraction and to assist in the fitting of dentures. Because of the difficulty in locating a reference plane within such a subject, the manoeuvrability of the contour planes offered by the sandwich technique is seen as being likely to succeed.

8.4.3. Prosthetic appliances

Examples of holographic studies in this area are quite numerous and have shown promise in measuring the wear of prosthetic devices and in optimising the shape of these appliances. Lalor *et al.* (p. 20 in ref. 49) have applied MRI holographic contouring principles to a number of prosthetic devices and Fig. 21 shows reconstructions from contourgrams of the medial section of a tribial component of a Manchester knee design worn *in*

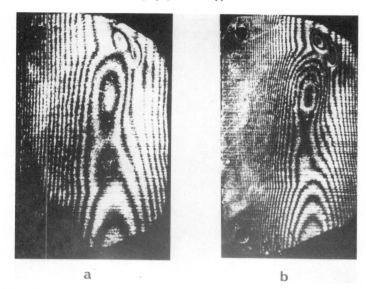

a   b

Fig. 21. Contourgrams of a worn Manchester knee, multiple refractive index method: (a) $113\,\mu$m contours; (b) $55\,\mu$m contours. (Courtesy of M. J. Lalor, Liverpool Polytechnic.)

vivo for about two years. The contour interval in Fig. 21(a) is $113\,\mu$m while in Fig. 21(b) it is $55\,\mu$m.

Høgmoen and Løkberg investigated the behaviour of an incusprosthesis (in the ossicular chain of the ear) to a range of audio frequencies to determine its response characteristics (p. 147 in ref. 36). The results of the investigation showed that the prosthesis behaved in a like manner to the real incus bone for frequencies up to 800 Hz, but above this value the vibration axis is rotated through $90°$. This condition is reversed again at frequencies approaching 2000 Hz and the authors suggest this is the likely explanation of reported cases of persons using the prosthesis losing their hearing in this frequency range.

8.4.4. Miscellaneous applications

Recently reported applications of holography to biomedical problems have included studies looking at the demineralisation of bones, fixation of fractures, testing of eardrums, detection of bladder tumours, viewing of neuroradiograms and a quantitative analysis of boundary layers in electrodialysis and ultrafiltration, to name but a few. The range of applications is thus seen to be quite extensive and new areas are being explored continually. It therefore seems highly probable that before long holography and holographic interferometry will become an integral part of routine medical investigations.

9. CONCLUSION

In the limited space available an attempt has been made to convey some impression of the wide scope of present and potential uses of the holographic technique; the reader can no doubt imagine many other situations in his own particular area of interest where these methods might prove to be of use.

In this rapidly expanding field it is almost impossible to keep abreast of all new developments and applications and apologies are rendered for having omitted any major new application which has been overlooked.

For further reading the list of journals following the references carry an abundance of papers on holography and its applications and are highly recommended.

REFERENCES

1. Gabor, D. *Nature*, **161** (1948), 777–8.
2. Leith, E. N. and Upatnieks, J. *J. Opt. Soc. Am.*, **52** (1962), 1123–30.
3. Leith, E. N. and Upatnieks, J. *J. Opt. Soc. Am.*, **53** (1963), 1377–81.
4. Leith, E. N. and Upatnieks, J. *J. Opt. Soc. Am.*, **54** (1964), 1295–301.
5. Powell, R. L. and Stetson, K. A. *J. Opt. Soc. Am.*, **55** (1965), 1593–8.
6. Burch, J. M. *J. Strain Analysis*, **9**(1) (1974), 1–3.
7. Abramson, N. *Proc. Conf. Engineering Uses of Coherent Optics*, ed. E. R. Robertson, 1976, Cambridge University Press, Cambridge, pp. 667–8.
8. Smith, H. M. (Ed.). *Topics in Applied Physics*, Vol. 20: Holographic Recording Materials, 1977, Springer-Verlag, Berlin.
9. Webster, J. M. Private communication.
10. Welford, W. T. *Appl. Optics*, **5** (1966), 872.
11. Thomson, B. I. *et al. Appl. Optics*, **6** (1967), 519.
12. Briones, R. A. *Proc. SPIE*, Vol. 215: Recent Advances in Holography, 1980, p. 112.
13. van Ligten, R. V. and Osterberg, H. *Nature*, **211** (1966), 282.
14. Collier, R. J. *et al. Optical Holography*, 1971, Academic Press, New York.
15. Nicholson, J. P. *et al. J. Phys. D: Appl. Physics*, **3** (1970), 1387.
16. Hildebrand, B. P. *Proc. Conf. Engineering Uses of Coherent Optics*, ed. E. R. Robertson, 1976, Cambridge University Press, Cambridge.

17. Burch, J. M. *Prod. Eng.*, **44** (1965), 431.
18. Brooks, R. E. *et al. Appl. Phys. Lett.*, **7** (1965), 248.
19. Ennos, A. E. *J. Phys. E. (Sci. Inst.)*, **1** (1968), 731.
20. Butters, J. N. *Proc. Conf. Engineering Uses of Holography*, eds E. R. Robertson and J. M. Harvey, 1970, Cambridge University Press, Cambridge.
21. Luxmoore, A. R. and House, C. *Proc. Int. Symp. Holography, International Commission of Optics*, 1970, Besançon, paper 5-2.
22. Boone, P. M. *Optics & Laser Tech.*, **2** (1970), 94.
23. Boone, P. M. *Proc. Conf. Engineering Uses of Coherent Optics*, ed. E. R. Robertson, 1976, Cambridge University Press, Cambridge.
24. Boone, P. M. *Optica Acta*, **22** (1975), 579.
25. Ennos, A. E. and Virdee, M. S. *Proc. IUTAM Symp. Optical Methods in Mechanics of Solids*, 1981, Poitiers, France.
26. Ennos, A. E. and Virdee, M. S. *Exp. Mech.*, **22** (1982), 202.
27. Dandliker, R. *Optics and Lasers in Engineering*, **1** (1980), 3.
28. Weir, R. *MSc Thesis*, University of Strathclyde, 1976.
29. Felske, A. and Happe, A. *Proc. Conf. Engineering Uses of Coherent Optics*, ed. E. R. Robertson, 1976, Cambridge University Press, Cambridge.
30. Felske, A. and Happe, A. *SAE Paper No. 780333*, 1979.
31. Erf, R. K. *et al. Proc. Conf. Engineering Uses of Coherent Optics*, ed. E. R. Robertson, 1976, Cambridge University Press, Cambridge.
32. Waddell, P. and White, D. *Jap. J. Appl. Physics*, **14**, supp. 14-1 (1975), 505.
33. King, W. *PhD Thesis*, University of Strathclyde, 1980.
34. Erf, R. K. (Ed). *Holographic Non-Destructive Testing*, 1974, Academic Press, New York.
35. Robertson, E. R. and Harvey, J. M. (Eds). *Proc. Conf. Engineering Uses of Holography*, 1970, Cambridge University Press, Cambridge.
36. Robertson, E. R. (Ed.), *Proc. Conf. Engineering Uses of Coherent Optics*, 1976, Cambridge University Press, Cambridge.
37. Robertson, E. R. and Elliot, S. B. *Proc. Conf. Applications of Holography*, Bescançon, 1970.
38. Abramson, N. *Proc. Conf. Engineering Uses of Coherent Optics*, ed. E. R. Robertson, 1976, Cambridge University Press, Cambridge, p. 631.
39. Dudderar, T. D. *Exp. Mechs.*, **9** (1969), 281.
40. Pflug, L. *7th Int. Cong. Italian Soc. for Stress Analysis (AIAS)*, Cagliari, 1979.
41. Sciammarella, C. A. *IUTUM Symp.*, Poitiers, 1979.
42. Vest, C. M. *Int. J. Nondestruct. Testing*, **3** (1972), 351.
43. Cadoret, G. *1st European Cong. on Optics Applied to Metrology*, SPIE, Vol. 136, 1977, p. 114.
44. Kapur, D. N. and Macleod, N. *Proc. Conf. Engineering Uses of Coherent Optics*, ed. E. R. Robertson, 1976, Cambridge University Press, Cambridge, p. 615.
45. Fraser, S. M. and Kinloch, K. A. R. *Laser + Elektro-Optik*, **7** (1975), 30.
46. Guerry, J. *et al. 2nd European Cong. on Optics*, SPIE, 1979, Vol. 210.
47. Greguss, P. (Ed.). *Holography in Medicine*, 1975, IPC Science and Technology Press, London.
48. Wolbarsht, M. L. (Ed.). *Laser Applications in Medicine and Biology*, Vol. 2, 1974, Plenum Press, New York.

49. von Bally, G. (Ed.). *Holography in Medicine and Biology*, Springer Series in Optical Sciences No. 18, 1979, Springer-Verlag, Berlin.
50. Greguss, P. *Opt. and Laser Tech.*, **8** (1976), 153.
51. Wuerker, R. *Proc. Conf. Engineering Uses of Coherent Optics*, ed. E. R. Robertson, 1976, Cambridge University Press, Cambridge, p. 533.

Further information concerning the application of holography can be found in the following journals:

Journal of The Optical Society of America; Journal of Physics, E, Scientific Instruments; Proceedings of Society for Experimental Stress Analysis; Optics and Laser Technology; Applied Optics; Optics and Lasers in Engineering.

APPENDIX: THEORY OF HOLOGRAPHY

The intensity of a light wave can be shown to be proportional to the square of its amplitude. When light waves arrive at a photographic plate the image density formed depends on the total energy received and is thus proportional to the square of the amplitude also.

In making a hologram two sets of light waves arrive at the photo-plate, the object and the reference beam. Since these are derived from the same source, they have a constant phase relationship. At any point (x, y) on the plate the reference beam can be described as $a_r \exp(-i\alpha)$ where α is a function of x, y and time, t. Because the reference beam is definable geometrically the function α is not difficult to determine; however, as we shall see, it is not necessary for the present purposes for the function to be stated more precisely than this. Similarly the object reflects light on to the photo-plate and at the same point (x, y) the object beam can be described as $a_0 \exp(-i\beta)$; in this case β is a complicated function which is generally indefinable.

The two beams combine to give a resultant amplitude $A = a_r \exp(-i\alpha) + a_0 \exp(-i\beta)$ and this will form an image whose density is proportional to $|A|^2$. The square of the modulus of a complex number is also equal to the number multiplied by its conjugate. Using this device we can thus express the darkening of the plate as being proportional to $A \cdot A^*$:

$$A \cdot A^* = (a_r \exp(-i\alpha) + a_0 \exp(-i\beta))(a_r \exp(i\alpha) + a_0 \exp(i\beta))$$

$$= a_r^2 + a_0^2 + a_r a_0 \exp(-i\alpha)(\exp(-i\beta) + \exp[i(\beta - 2\alpha)])$$

In the reconstruction stage, the reference beam $a_r \exp(-i\alpha)$ is shone alone

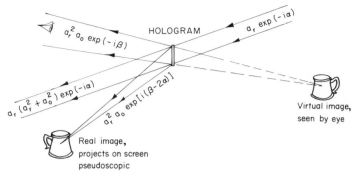

Fig. A1. How one small hologram copes with all that mathematics!

on to the processed plate. It is modulated by the information recorded on the plate to give an amplitude $\mathscr{A}$, where

$$\mathscr{A} = a_r \exp(-i\alpha)\{(a_r^2 + a_0^2) + a_r a_0 \exp(i\alpha)(\exp(-i\beta) + \exp[i(\beta - 2\alpha)])\}$$
$$= \exp(-i\alpha)a_r(a_r^2 + a_0^2) + a_r^2 a_0 \exp(-i\beta) + a_r^2 a_0 \exp[i(\beta - 2\alpha)]$$

The first of these terms simply indicates that a darkened version of the reference beam is diffracted through the plate, travelling in the same direction.

The second term $a_r^2 a_0 \exp(-i\beta)$ represents the original object beam $a_0 \exp(-i\beta)$ modified only in brightness by a_r^2. By putting more energy into the reconstruction reference beam we can easily reproduce the image as bright—or brighter—than before. It is important to notice that the descriptive part $a_0 \exp(-i\beta)$ is obtained exactly as before. It is a precise restatement of the original light reflected from the object and gives a virtual image of the object which can be seen by eye when looking through the hologram.

The third term in the above expression $a_r^2 a_0 \exp[i(\beta - 2\alpha)]$ represents another diffracted wave similar to the object beam but travelling in the opposite direction and making an angle to the hologram plate twice that of the original reference beam direction. This gives a real image of the object but with reversed sense of depth (it is pseudoscopic) and may be projected on to a screen. These three reconstructed terms are illustrated in Fig. A1.

Chapter 7

SPECKLE INTERFEROMETRY

J. N. BUTTERS

Department of Mechanical Engineering, Loughborough University of Technology, UK

1. INTRODUCTION

Speckle pattern is a phenomenon which has been studied for many years in specialist applications and the first statistics of speckle phenomenon to be published were probably those due to Lord Rayleigh.[1] The appreciation of speckle effect was restricted to particular aspects of astronomy before the advent of the laser and its capability of producing highly coherent light.

A characteristic of laser illumination is the frosty appearance which it gives to most surfaces. This appearance is the speckle effect and the size of the apparent grains is dependent on the viewing optics, often the naked eye. In the early days of holography this speckle effect was talked of in the general sense of noise and attempts were made to remove it since it degraded the appearance of holographically formed images.

When interpreting a holographic interferogram the direction cosines of the illuminating and viewing direction must be defined. If one sets up viewing equipment for such interpretations, then in order to define the direction cosines practically, both the viewing and illuminating apertures have to be made small. This in turn causes a strong enhancement of the speckle effect which seriously degrades the image being viewed and it was this problem which initiated work, carried out by Leendertz,[2] into using speckle pattern for measurement. It is now recognised that a single speckle is itself a form of interferogram; in the live sense it has amplitude and phase associated with it. From the information point of view, this is all that is

205

necessary to define a position in space of the scattering surface which is responsible for forming the speckle.

The application of speckle to metrology is really a matter of determining an optical system which will enable the data encoded into the speckle to be extracted. The possibilities have given rise to many experimental and theoretical investigations, the main aspects of which are presented in this chapter, although of necessity, individual special details are omitted, unless they are in regular use. The majority of workers have concentrated on the processing of speckle recorded in photographic negatives, the processing of such negatives being similar in many respects to classical moiré fringe work. The phase content of the speckle is then fixed by the recording method. The approach at Loughborough University has always been to use the phase encoded into the speckle directly, which is achieved by the use of a reference beam, as in holography. If a speckle pattern is formed using an imaging system from the light scattered by an optically rough surface on to a plane which also contains the phase invariant wavefront, then the speckles are redistributed according to a known phase distribution. Movement of the scattering surface will result in phase changes within the coherent image and these phase changes will, in turn, when integrated with a reference beam, give variation in the point to point brightness of the speckle pattern. If this brightness variation is monitored, then points of equal movement of the scattering surface will be highlighted in terms of interference fringes and in fact the system behaves almost identically to the fairly well known process of real time holographic interferometry. There is one significant practical difference: in order to record a hologram, one needs a photographic emulsion resolution capable of recording a diffraction grating and often approaching 3000 line pairs per millimetre. The holographic process depends on the reconstruction of wavefronts which, in turn, interfere to form the interference patterns in holographic interferometry. In the case of the speckle methods, no reconstruction of wavefront is required and the resolution of the recording material need only match the speckle size in the image. This speckle size can be varied by choosing the aperture of the optical recording system and there is the possibility of matching the speckle size to the resolution of a television camera (which forms the basis of the electronic speckle pattern interferometer (ESPI)).

Several versions of this interferometer are now in existence: in Europe, the US and UK. The instrument described in this chapter will be that due to the researchers at Loughborough University. The advantages of using an ESPI are that the rapid recording of interferograms is possible and that data processing can be conducted on line, the result being displayed in analogue form on the monitor at the standard television rate. The data contained in the interference fringes is in analogue electronic form and it is possible to interface this directly to the computer. In the past this has been done in a fairly limited form because of the high data content and therefore large storage requirements of the computer, but now with increasing capability per unit of computer cost, it is feasible to think of AD conversion of the image and digital processing, possibly with a re-display of the final data.

With holographic interferometry, actual wavefronts are recorded which are influenced by all changes in the object position. The interferogram is formed at the interpretation stage by the interference of wavefronts which require analysis on a full spatial basis. With speckle pattern work sensitivity can often be restricted to a chosen parameter or motion direction. In the case of speckle photography in-plane motion or tilt can easily be measured separately under ideal conditions whilst with speckle pattern inter-ferometry in-plane motion and out of plane motion can be separately identified but tilt is less easily measured as a single parameter.

When using an ESPI type instrument, it is possible, by using two wavelengths of laser light, to produce shape difference contours. This enables speckle pattern interferometry to be applied as a comparator in the true sense, the actual contour interval defining the shape difference resolution and being given by half the product of the two laser wavelengths involved divided by their difference $[\lambda_1 \lambda_2 / 2(\lambda_1 - \lambda_2)]$. Once set up with either conventional optical or holographic means of generating a reference wavefront, the apparatus can inspect components optically at the rate of 10 per second, although in practice the speed is very much less than this being limited by the rate at which the components to be inspected can be fed into the machine. In order to accomplish this form of optical comparison it is necessary to separate the effects of macroscopic shape from microscopic surface detail. As a comparator only macroscopic shape data is used, but it is equally possible to record the amount of speckle decorrelation between the two wavelengths used which provides a measure of surface roughness detail.

The actual application of speckle pattern methods is very wide indeed. The most obvious applications are in laboratory and development areas where deformation is measured either quantitatively or qualitatively. An example of quantitative measurement is in strain determination, but qualitative data is often sufficient for NDT.

The choice of photographic or other methods will be made according to circumstances, the relative merits being outlined in the text and summarised in the last section. In some cases a comparative measurement must be carried on over a long period of time and in this case a form of instrument which can refer to a standard would be preferred. Similar systems could be chosen where variation in surface finish would affect speckle formation since the two wavelength systems can measure shape with considerable independence of surface roughness.

Finally, speckle methods can be used to measure surface roughness. Many techniques exist, some of them having considerable promise, but none accord with existing engineering standards of surface finish based on stylus instruments. Fundamentally it would seem appropriate to measure an area based quantity by an area based technique rather than to express a quantity with dimensions L^2 in terms of a linear expression in L.

2.   BASIC CONCEPTS USED IN SPECKLE METROLOGY

In most cases the speckle pattern utilised will be affected by the influence of a lens somewhere in the optical recording system which is often an ordinary film or television camera. The actual speckle size processed is then generally different to the speckle size formed by diffraction at a circular aperture given by the Airy disc diameter, S_1

$$S_1 = 1 \cdot 22 \lambda f \tag{1}$$

where λ = wavelength and f = 'f' number of the system. When the speckle pattern is focused by an optical system of magnification M the speckle diameter seen by the film or television vidicon tube is S_2

$$S_2 = 1 \cdot 22 \lambda f (1 + M) \tag{2}$$

It is sometimes necessary to consider the effective speckle size on the surface of an object for a given recorded speckle size S_2. If this assumed surface speckle is of diameter S, then by definition:

$$S_2 = SM$$
$$S = 1 \cdot 22 \lambda f \left(\frac{1 + M}{M} \right) \tag{3}$$

This places limits on the measurement sensitivity range for very large and very small objects.

The first known work on speckle interferometry was conducted by Leendertz[2] at Loughborough University. It was first necessary to establish

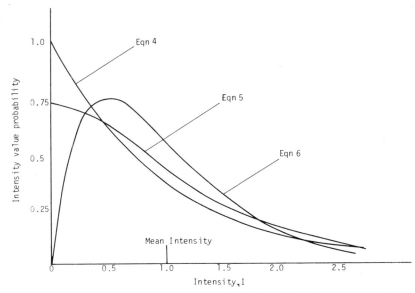

Fig. 1. Curves showing the probability density functions for common speckle pattern distribution conditions.[4] The similarity of the distributions for I above 0·3 is important to the range of applications for speckle pattern interferometry.

that the speckle pattern was of adequate contrast and it was expressed that the speckle intensity followed a negative exponential relationship. Whilst Leendertz's work considered the addition of intensities from two speckle patterns, speckle pattern interferometry usually operates by combining a speckle image with a wavefront of constant phase relationship. The work of Leendertz follows that of Rayleigh[1] but further studies of combined speckle fields applicable to coherent speckle metrology have been conducted by Goodman.[3]

In summarising speckle effects observed in metrology, work by Ennos[4] has produced curves for the probability of a grain speckle intensity plotted against the average intensity value for three cases: the single speckle field

$$P(I) = \frac{1}{I_0} \exp\left(\frac{-1}{I_0}\right) \tag{4}$$

the combination of a speckle with a plane wave:

$$P(I) = \frac{2}{I_0} \exp\left[-\left(1 + \frac{2I}{I_0}\right)\right] J_0\left(2\sqrt{\frac{2I}{I_0}}\right) \tag{5}$$

and the incoherent combination of two speckle patterns

$$P(I) = 4\left(\frac{I}{I_0^2}\right)\exp\left(-\frac{2I}{I_0}\right) \tag{6}$$

These three statistical patterns seem quite different but over the main working range there is considerable similarity, as shown in Fig. 1.

3. SPECKLE PATTERN INTERFEROMETRY AND SPECKLE PHOTOGRAPHY

A much more significant variation to be found in speckle methods lies in the interpretation of measurement data for the recordings. In this respect there is divergence of opinion as to what speckle interferometry really is. In the author's opinion interferometry describes a process where two coherent, or partially coherent, fields of light are brought together. This would include methods where two speckle fields are combined or a speckle field is combined with a reference wave as in holography. This definition would not include many techniques of speckle photography considered later, which are closely associated with moiré methods. However, taking the broader view, speckles are formed by interference and both techniques can be applied to many measurements.

4. IN-PLANE DEFORMATION

The original work by Leendertz,[2] considered the addition of two speckle fields and the method was experimentally demonstrated using two speckle fields generated from a single scattering surface using the illumination arrangement shown in Fig. 2. It is likely that the surfaces used de-polarised the light to some degree, resulting in intensity distributions according to eqn (6), rather than the negative exponential law assumed, but the success of the experiment demonstrates the small practical effect of the statistical differences. Using the notation in Fig. 2, it may be expected that the speckle pattern in the image plane would be unaffected by movements in the Y or Z directions, but a movement of x in the direction of X would change the phase by

$$\Delta\lambda = \frac{2\pi}{\lambda}\cos\theta\,[x - (-x)]$$

$$= \frac{4\pi x}{\lambda}\cos\theta \tag{7}$$

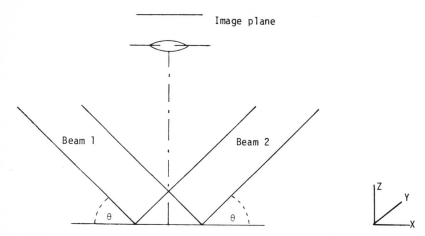

Fig. 2(a). Beam arrangement for measurement of in-plane displacement by speckle pattern interferometry.

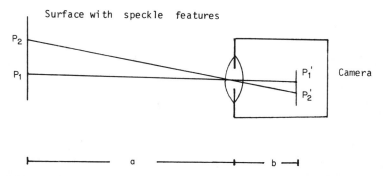

Fig. 2(b). An arrangement for speckle photography made sensitive to in-plane surface movement. Points such as P_1–P_2 on the surface are imaged on the film plane. In most applications distance $a = b$, but for precision strain measurements a is reduced. Maximum accuracy is obtained for unit magnification when $a + b = 4F$. (F is the focal length of the lens.)

A uniform movement of the surface will result in a cyclic variation of average brightness in the image plane. Non-uniform motion, however, for example that produced by elastic strain, will result in differing phase changes within the field of view since x is then variable. This effect will show as a fringe pattern. The device can be used as an optical strain gauge providing that fixed positions can be identified by, say, surface marking. The fringe pattern will enable changes in length to be calculated according to eqn (7) and the quantities defining strain e

$$e = \frac{\text{change in length}}{\text{original length}} \tag{8}$$

may be determined as follows.

Let m be the number of fringes between two marked points, distance l apart. The change in l is then determined according to the equation

$$2m\pi = \frac{4\pi}{\lambda} x \cos \theta \tag{9}$$

or

$$x = \frac{m\lambda}{2 \cos \theta}$$

$$e = \frac{x}{l \pm x}$$

usually $l \gg x$ thus e may be expressed by x/l; substituting

$$e = \frac{m\lambda}{2l \cos \theta} \tag{10}$$

A complete instrument for measuring all surface components of strain has been produced by Jones[5] at Loughborough. The process does have limits due to polarisation and correlation effects in a practical instrument. It is usually possible to obtain X sensitivity better than ten times the Y or Z sensitivity and a practical range is approximately $100 \, \mu$m strain to $1000 \, \mu$m strain. It is, of course, possible to use the instrument repeatedly over a large strain range, this being so when using an ESPI based instrument, later described. The instrument is particularly beneficial for viewing strain fields. Figure 3 shows one frame from a video tape recording by Denby at Loughborough, of the strain effects in a carbon fibre reinforced epoxy strip with a hole and subjected to an increasing tensile load. This demonstrates very forcibly the value of seeing a strain field. Attempts to use the

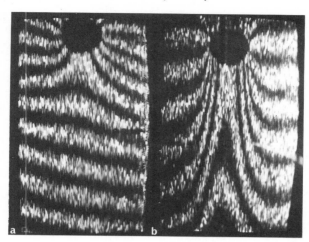

Fig. 3. Comparison of strain fields for an illumination strip (a) and a carbon fibre reinforced plastic strip (b) both subject to tensile load. Visualisation carried out with ESPI and plates taken from videotaped record of dynamic results.

instrument for examining very small strain fields, as for example at crack tips, have unfortunately not been generally beneficial to date since the magnification effects on speckle size, shown in eqns (2) and (3), together with surface roughness effects, limit the optical magnification that can be used. Experiments at Loughborough University showed a limiting value of M of about 16, with existing image processing methods. These particular size limitations are not relevant for large surfaces where M is less than 1. Practical limits with larger components are usually related to component stability and the necessary compensation for body motion to avoid speckle decorrelation.

When the interferometer is set to an operational condition, the process described is reliable and in an *ad hoc* calibration test using a wire strain gauge, the agreement between the optical strain measurement and the electric strain gauge measurement fell within 2 % over the whole range. The optical instrument tends always to fail safe since if working conditions are not satisfied, fringes are not generally observed. The difficulties of operating this form of interferometer has encouraged a number of workers to turn to photographic speckle techniques which show considerable promise for strain measurement. In this latter case results are usually obtained, but uncertainties in their meaning led to an investigation by Archbold and others,[6] who concluded that accurate strain determination by this method demanded severe constraints on the experimental arrangement. For an experiment cited in ref. (6), errors of 300 microstrain were possible due to both spurious 'line of sight' motion and tilt, and an error of 120 microstrain was an observed possibility due to optical aberrations of a telecentric|viewing|system used. In addition to this, surface stability is necessary to avoid decorrelation effects which become serious when surface movements exceed the speckle size. A conclusion reached in ref. (6) was that whilst speckle photography has the advantages of convenience for in-plane movement measurements, probable errors are considerable in high sensitivity work such as strain measurement. Quoting from the paper 'even under the best conditions of a flat object imaged by a well corrected lens of limited field, uncertainties of about 100 microstrain may still occur'.[6]

5.   SPECKLE PATTERN IN VIBRATION STUDIES

Vibration mode visualisation has, since the early days of holographic interferometry, been recognised as one service which is performed more elegantly by laser light than by any other means. In practical cases it is usual for the distance between the vibrating component and the plane of recording of the optical field to be large compared to the amplitude of vibration. Let the vibration be described by:

$$z(x, t) = Z(x) \sin \omega t \tag{11}$$

The scattered light leaving the vibrating surface then has the form

$$U_S(x, y) = A(x; y) \exp [i\phi(x, y)]$$

Since the optical path change is twice the motion amplitude, the resulting phase change due to vibration denoted by $\Delta\phi$ is

$$\Delta\phi(x; y; t) = \frac{2\pi}{\lambda} 2Z(x) \sin \omega t$$

$$U_v(x. y, t) = A(x; y) \exp \left\{ i \left[\phi(x, y) + \frac{4\pi}{\lambda} Z(x) \sin \omega t \right] \right\}$$

For either holographic work or speckle interferometry, where the recording period is greater than several cycles of the vibration, U_V is added to a reference wave and integrated over the recording period T.

$$I_{(t)} = \int_0^T (U_V + U_R)\,dt \qquad U_R(x, y) = A_R(x, y) \exp\left[i\phi_R(x, y)\right]$$

$$I_{(t)} = A_r^2 + A_c^2 + 2A_r A \cos\{\phi_R - \phi\} J_0 \frac{4\pi}{\lambda} Z(x) \tag{12}$$

The contrast in the interference pattern is controlled by the last term showing a periodicity of

$$J_0^2 = \frac{2\pi}{\lambda}\,(x) \tag{13}$$

having a maximum for $x = 0$ and minimum $x = m\lambda/4$. The fringe intervals occur at the Bessel function zeroes which, for the lower orders, are not equally spaced. In holographic interferometry many fringes are often visible but with speckle pattern work only the first few fringes will be seen making the variable spacing important in any quantitative interpretation. Values of the displacements shown are given in Table 1.

The types of apparatus used for observing vibration modes by speckle pattern are very varied. In the simplest case viewing a laser lit surface with the naked eye or the eye assisted using a card pricked with a small hole, will enable nodes to be identified. Some people find this method difficult, but most are successful after practice. The method is restricted to identifying

TABLE 1

Actual Vibration Amplitudes at the Centres of the First Seven Fringes Obtained by the Time-Averaged Method of ESPI Assuming Green Laser Light at $0.5\,\mu m$ Wavelength

Fringe number	Bessel function zero value	Deflection (μm) from static position for $\lambda = 0.5\,\mu m$	Motion amplitude (μm) for $\lambda = 0.5\,\mu m$	Error (μm) assuming fringes at $0.5\lambda = 0.25\,\mu m$
1	2.4048	0.096	0.191	0.059
2	5.5200	0.220	0.440	0.06
3	8.6537	0.344	0.689	0.061
4	11.7915	0.469	0.939	0.061
5	14.9309	0.594	1.189	0.061
6	18.0710	0.719	1.438	0.061
7	21.2116	0.844	1.688	0.062

nodes only and no fringes are visible due to the restricted response of
human eyes.

A neat and simple interferometer which assisted viewing was prepared by
Stetson during a period at the National Physical Laboratory (NPL), the
principle of which is shown in Fig. 4. The scattered light from the object is
combined with a reference beam in an instrument rather like a modified
telescope. The reference beam is introduced by a beamsplitter at
approximately the image plane of the object, this beamsplitter being of
wedge form to prevent its action as a Fabry-Perot interference generating a
fringe pattern visible through the eyepiece. An iris diaphragm is positioned
in front of the objective lens to adjust its aperture and thus regulate the
speckle size giving the best fringe visibility to the viewer. The instrument is
easy to use in the laboratory, but requires a low level of background light
and, as with all speckle interferometers, there is a need to balance reference
and object beam light path lengths to well within the coherence length of the
laser. With most HeNe lasers this tolerance is approximately $\pm 30\,\text{mm}$

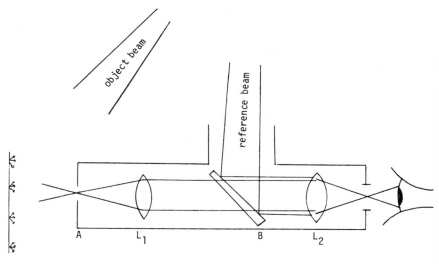

Fig. 4. Diagrammatic arrangement of Stetson's speckle interferometer. The
reference and object beams must originate from the same laser and have path
lengths compensated to the laser coherence length. The lenses L_1 and L_2 need to be
well corrected for spherical aberration. To improve speckle contrast when viewing a
depolarising surface, a polaroid may be placed in front of lens L_1. The instrument
inverts the image of the viewed surface, correction for this being possible by placing
an Amici prism in front of the iris diaphragm, A.

although this extends to 500 mm or more for argon lasers fitted with mode selecting etalons.

The simplicity of vibration viewing arrangements, led to the choice of this application for the development of ESPI in the author's laboratory at Loughborough. At the time a number of observations were being made of the resonant mode shapes of engineering components using time-averaged holographic interferometry. The large number of holograms required for this work encouraged the search for a television based method for obtaining the same vibration data. Speckle offered this possibility since its spatial content could be confined within the resolution capability of a television camera and this provided the obvious motivation for Leendertz's work. It should be remembered, however, that to obtain a speckle pattern an aperture has to be introduced into the system recording the light wave scattered from the object which actually reduces the resolution available in the image. Whilst ESPI, therefore, is an extremely convenient way of viewing vibration modes the clarity of the interferogram is reduced by comparison with time-averaged holographic interferometry. The opportunity for frequency filtering the video signal and an associated selective amplification, however, enables the speckle pattern interferogram to be displayed with much greater clarity and showing more fringes than are possible with an optical speckle arrangement alone. Even so the dynamic range of photographic materials is much larger than for a television vidicon where the effect of the J_0^2 function in the fringe intensity distribution forms a very much more noticeable limit on ESPI methods than on holographic methods. Using a flat vibrating scattering surface for reference, a vidicon system is limited to seeing seven amplitude of vibration mapping fringes by the time-averaged method whereas holographically, using the same process, 50 or more such fringes may be viewed. This may initially appear to be a serious limitation for speckle, but in practice amplitude information is usually irrelevant and the fringe pattern is used only to aid determination of nodal positions with some precision.

There is a range of vibration mode determination requirements where vibration amplitude is substantially less than the wavelength of light. This problem has been treated holographically by phase modulating the reference beam, the same process being applicable to speckle pattern interferometry. Phase modulation may be introduced by movement of a reference beam mirror which in this case would be formed on a piezoelectric substrate. In the author's laboratory, Cookson and Koyuncu[7] have transmitted the reference beam through a KDP electro-optical modulator crystal which provided extremely reliable performance with fringe sensitivity below 0·01 μm. The application in this case related to the study of the vibration modes of ultrasonic transmitters.

6. SPECKLE PHOTOGRAPHY

Speckle photography has been applied to the largest size range of measurement tasks and can be adapted to sense all forms of motion that can be identified by looking at an object or surface either by eye directly or with the assistance of an instrument. The attractions of the technique are its simplicity, its sensitivity range and relatively low cost of apparatus. Even white light speckle pattern interferometry has been carried out by some workers, Burch,[8,9] in particular, having used the dot pattern from printer's half-tone as a surface masking medium and also the surface structure itself when working with woven fabrics. Requirements for mechanical stability are of the order of the recorded feature size, be that either a coherent speckle or printed dot, making the technique far less demanding of environmental conditions than pure interferometric studies either by holography or speckle. There are two principal methods of photographing speckle effects which give independent recordings respectively of in-plane displacement and tilt. Most practical speckle photography is not carried out under ideal conditions so that recorded changes may be due in part to more than one motion. This does cause a problem, but it is appropriate to appraise the ideal conditions first.

6.1. In-Plane Speckle Photography

For the measurement of the in-plane displacement of a macroscopically flat, rigid surface the arrangement shown in Fig. 2(b) is employed. This may be compared with the interferometric arrangement shown in Fig. 2(a) which actually quantifies displacement in approximately half wavelength intervals.

The optical system is adjusted to accurately focus an image of the coherently lit surface onto the recording plane. This image is modulated by a speckle pattern of the size given by eqn (1), $S = 1·22\lambda f(1 + M)$, where 'f' is the 'f' number of the lens system. The actual formation of the speckle pattern will depend on the scattering characteristics of the surface which will not change as a result of a lateral translation with respect to the viewing axis. The image at the recording plane will thus translate in position in exact proportion to the surface movement, the constant of proportionality being the system magnification M. The recording is independent of the illuminating direction although this should preferably remain constant during the practical experiment.

Measurement of an in-plane displacement, say in the direction of Y is achieved by recording the speckle pattern twice, once before the translation and again afterwards, both recordings being on the same piece of film which is held fixed throughout the test. If the movement in the Y direction, assumed greater than a speckle size, is denoted by y, then the film will have two identical speckle patterns recorded upon it, distance $y' = My$ apart. Knowing the system magnification M, the translation y is easily obtained from the measurement of y'. In practice y' would normally be obtained by simple optical processing of the film using apparatus in the form of a linear diffractometer (Fig. 5). The converging spherical wave of coherent light formed by the transforming lens has a diffracted component introduced by the speckle. The regular displacement y' of the speckle acts rather like a grating giving a well-defined fringe pattern in the focal plane. The spread of light in the focal plane is often called the speckle halo and consists of a central bright spot of undiffracted light surrounded by another speckle pattern formed by the random diffraction effect from the illuminated area of the speckle recording. This new speckle pattern will however be crossed by a series of fringes which can be well-spaced and which give a direct measure of the recorded displacement y', an example of such a pattern by Jones is shown in Fig. 6. The value of y' may be obtained by considering the fringes to be formed effectively by a Young's double slit arrangement (Fig. 7) where using the notation shown on the figure:

$$d = \frac{\lambda l_1}{y'} \qquad \text{thus } y = \frac{\lambda l_1}{Md} \text{ (since } y' = My)$$

The fringes seen in the image plane of the diffractometer lie in a direction normal to the displacement vector. In this case for a displacement entirely in the Y direction the fringes lie parallel to the axis of X. In general measured displacements will involve both X and Y contributions which are obtained separately by resolving the vector displacement indicated by the fringe pattern.

Uniform in-plane translations represent only a trivial case of practical application. A common requirement is to measure strain, defined earlier in eqn (8). Clearly a surface undergoing strain will exhibit different displacement values at differing locations on the surface. The interpretation of the speckle data involves determination of displacement as a function of position. This can be achieved fairly easily from the doubly

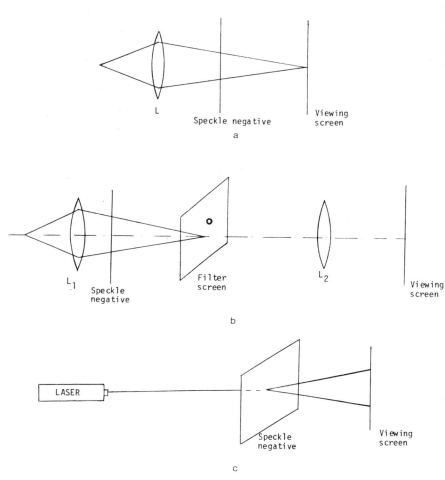

Fig. 5. (a) Arrangements for obtaining fringe data from doubly exposed speckle negatives. A simple system suited to measurement of whole body motion. (b) A development of system (a) using a filtering screen and imaging lens L_2 to provide an interferogram indicating localised displacement between exposures. (c) A simple method for observing differential displacements over a component surface for the doubly-exposed speckle negative.

Fig. 6. Typical fringes obtained from the speckle halo formed by passing an unexpanded laser beam through a doubly-exposed focused image speckle negative. A small in-plane displacement of the scattering surface was made between exposures.

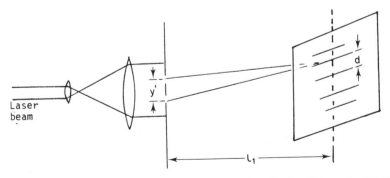

Fig. 7. A Young's double slit arrangement where the effective slits are formed by speckle pairs imaged on the speckle negative distance y' apart.

exposed speckle photograph, either by using additional components in the diffractometer, as shown in Fig. 5(b), or by simply scanning the pattern with the unexpanded beam from a low power laser (see Fig. 5(c)). In either case the fringe pattern displayed on the viewing screen relates to the displacement recorded in the region (x, y) through which the light is passed. This therefore enables a displacement map to be drawn from which strain can be calculated. Some practical problems associated with this method relate to errors arising from optical aberrations and also to the fact that most components of interest are not accurately represented by a plane surface moving normal to the viewing axis. Curvature of the surface leads to error due to lack of focus allowing sensitivity to tilt and movement in the direction of line of sight. The latter effect is generally small, affecting the speckle through variation of magnification due to change in object position.

6.2. Tilt Measurement by Speckle Photography

Measurement of the tilt of a macroscopically flat plane surface is obtained by processes similar to those obtained for translation with one important difference. The camera is now focused to infinity so that only parallel light is brought to a focus at the film plane. Since the surface detail is no longer imaged the speckle pattern is formed from light scattered from any point on the surface into the selected direction. Translation of the scattering surface in its own plane does not therefore result in any change in the recorded speckle pattern. A small tilt in the scattering plane will result in a change in the position of the speckle in the image plane because the camera has been set to discriminate ray direction. Using the notation of Fig. 8, if the surface moves from position AB to AB' through a small angle β the Z movement of a position y on $AB = \beta y$.

The change of optical path length resulting from this is $\beta y(1 + \cos \theta)$ and the speckle phase will change by

$$\Delta\phi = \frac{2\pi}{\lambda} \beta y(1 + \cos \theta) \tag{14}$$

For a lens system set to focus parallel light, the focal point of a parallel beam incident at angle γ to the optical axes of the lens is given by

$$d = F \tan \gamma \qquad \text{(or } F\gamma \text{ for small angles)}$$

The linear phase variation with y can be equated to the angle of movement

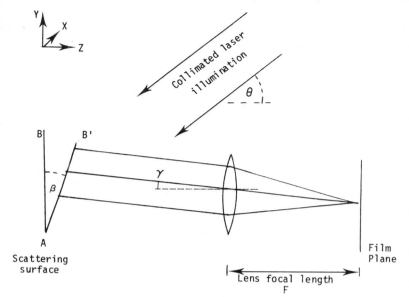

Fig. 8. Arrangement for recording the double exposure speckle negative to measure surface tilt indicated by above.

of the incident parallel beam to produce the same effect. The angle term is $(1 + \cos\theta)\beta$. Substituting to obtain an equivalent speckle displacement, d

$$d = F(1 + \cos\theta)\beta \qquad (15)$$

Using this relationship between tilt angle and speckle displacement the measurement is carried out by taking two exposures as before, and then optically processing the recording in the diffractometer. The separation of fringes seen in the diffractometer, d_f, can be used to obtain the tilt angle

$$\beta = \frac{\lambda F_T}{F(1 + \cos\theta)d_f} \qquad (16)$$

Variation of tilt cannot be easily measured by this method since light from the whole illuminated surface is integrated by the camera. Repeat of the test for different areas of illumination is, of course, possible but then provision must be made to relate each test to the same axis of reference.

7.   PRACTICAL EXPERIENCES FOR SPECKLE PHOTOGRAPHY

For industrial measurement the ideal cases for the measurement of in-plane deformation or tilt are seldom met and it is normal to record proportions of each effect in applied speckle recording. Gregory[10] has studied this problem in considerable detail. For the purpose of analysis each scattering centre is considered as an elemental mirror when ray diagrams may be used to interpret the recorded speckle. In particular Gregory has shown that both tilt and translation can be derived from a single set of double exposure speckle photographs obtained from a surface illuminated with divergent laser light. Assumptions are made that translation components are independent of tilt and illumination direction when the camera is accurately focused on the real surface. Gregory observed that tilt data could be obtained by focusing the camera on the plane of the virtual image of the source formed by the mirror facets, this representing a new, although empirical development on the subject. Stetson[11] has mathematically analysed the problem of speckle recordings which contain both tilt and displacement data and has shown that the information can be separated if two recordings are taken from the same position with arbitrary, but differing planes of focus. This work goes on to show that strains and rotations can be extracted if the test is repeated for three separate camera positions. Gregory has used speckle methods to show how features such as bulges and depressions can be determined by speckle photography, the main intention for the examples quoted being non-destructive testing. In one case permanent deformation curves along a crack for a high pressurisation cycle are given with a quantitative scale of depression expressed in microns. Such highly quantitative data are perhaps a little unexpected from the method. The direct sensitivity to out of plane or line of sight motions is low for speckle photography relying on the change in magnification M for changing surface position. Sensitivity is thus of the order of 1 mm at best, often leaving it necessary to calculate displacement from the measured angular tilt.

8.   ELECTRONIC SPECKLE PATTERN INTERFEROMETRY

The techniques of ESPI involve the use of two concurrent, separate beams except in the one example of surface roughness assessment. The apparatus can therefore be described as a two-beam interferometer having a basic sensitivity related to the wavelength of light. In this respect it has the same sensitivity as holographic methods but, due to the essential restricted aperture, fringe sharpness is less than in holography, necessitating a limit of fewer fringes in the field of view. Since the fringe display in ESPI work is generated from the relative position of focused speckles, the fringe pattern is always localised which removes the ambiguity often associated with the interpretation of fringes in holographic interferometry. The interferogram produced is displayed on a television monitor but obviously, since the data is in electronic form, it is possible to interface directly to a data processor. The measurement of out of plane motion using speckle pattern interferometry has, so far in this text, received very little mention. This is one function which is easily carried out with ESPI and will be used to describe the operation.

Speckle formation is often described by the combination of light vectors scattered from a large number of points on a scattering surface and being brought together at a point. In ESPI the resulting light amplitude vector from this combination is then itself combined with a reference beam to give phase sensitivity to the recording as in holography. Expressing the two contributions at point (x, y) as $U_0(x, y)$; $U_R(x, y)$,

$$U_0(x, y) = A_0 \exp i \left(\phi_0 + \frac{2\pi l_0}{\lambda} \right)$$

$$U_R(x, y) = A_R \exp i \left(\phi_R + \frac{2\pi l_R}{\lambda} \right)$$

the combined intensity

$$I(x, y) = A_0^2 + A_R^2 + 2A_R A_0 \cos \left[\phi + \frac{2\pi}{\lambda} (l_0 - l_R) \right] \quad (17)$$

where ϕ is the phase difference $(\phi_0 - \phi_R)$. When this intensity distribution is seen by the television camera the constant term can be suppressed leaving the variable cosine term which varies between unity and zero depending upon the difference $(l_0 - l_R)$ which are respectively the optical path lengths from the beam splitter to point (x, y) for the object and reference beams. Clearly the process is cyclic for values $(l_0 - l_R) = n\lambda$ where n is an integer. In practice l_R is usually fixed and l_0 is varied by surface movement. For a scattering surface, a surface movement of d would introduce a change in l_0 of $2d$ for normal viewing and a quantity close to this for small off-axis viewing angles.

To operate ESPI a video recording of the speckle pattern is made for both l_0 and l_R fixed. The most convenient recording method is to use a single television frame video disc recorder which will record the above speckle pattern, now referred to as the reference pattern, in one television frame period or 40 ms on the British and Continental standards. This picture is then electronically inverted in the sense that black speckles are displayed as white. The resulting signal is then fed into an electronic summer together with the live picture from the camera, the combined signal being displayed on the monitor. If the value of $(l_0 - l_R)$ is not changed between recording the reference pattern and subsequent viewing, the processed signal from the recorder should be the exact negative of the real time picture from the camera. This condition may be used to set up the ESPI apparatus, some adjustment usually being necessary to correct for different signal timings from the camera and the recorder. When the two patterns are brought into exact register the monitor screen should quite suddenly go dark. When this setting is reached the apparatus is ready to observe out of plane components of motion of the surface under investigation on an interferometric scale. The real time interferogram will be shown on the monitor and can be conveniently recorded on a standard video tape recorder. Examination of any particular feature can then easily be carried out subsequently using the still frame facility on the video tape recorder.

8.1. Vibration Studies by ESPI

The integration process involved when using time-averaged methods is given in Section 5 together with reference to the use of ESPI in time-averaged work. When setting up for this application it may be worth considering the use of off-axis illumination such as used with holographic interferometry. Figure 9 shows the two common arrangements. The off-axis system (Fig. 9(a)) has an advantage in that the lens diaphragm is the principal aperture restriction in the system, whereas in Fig. 9(b), the two folding mirrors form a frequency limiting condition which, in some cases, reduces fringe clarity. The arrangement of Fig. 9(b) is, however, more economical in its use of light and there is less loss of detail due to shadows when on-axis illumination is used on other than plane surfaces. The arrangement in Fig. 9(b) is necessary when using the instrument for two-wavelength contouring and thus it tends to be adopted for other applications, even though it may not always be advantageous.

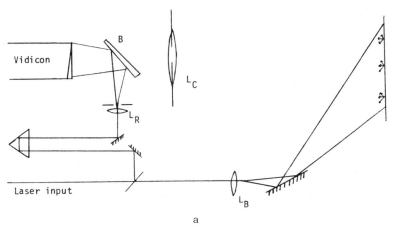

a

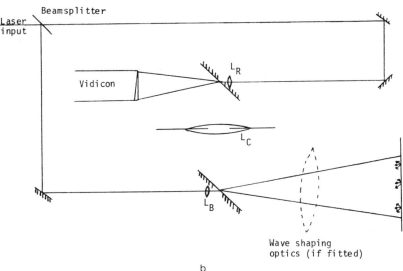

b

Fig. 9. (a) An off-axis arrangement of ESPI. L_C = camera lens, L_R = reference beam lens, L_B = lens for expanding beam for object illumination. This type of layout is particularly good for vibration studies using the time-averaged method. (b) A general form of ESPI but avoiding beamsplitter B by use of a folded mirror arrangement with central holes in the mirrors to transmit beams at the focal diameters of L_R and L_B. This 'normal viewing' arrangement avoids shadow and is particularly suitable for optical comparator applications.

The main principles to be considered in setting up a practical system, as in Fig. 9(a), are

(i)   the reference beam path length is adjusted to be equal to the mean object beam path length, incorporating a folded optical path if necessary for the reference beam; (whole laser cavity length variations are, of course, permitted);

(ii)   a wedge beam splitter is used to avoid spurious interference fringes in the field of view;

(iii)   the focal point of the reference beam lens, L_R, is arranged to coincide optically with the aperture of the camera lens, L_C, to minimise the frequency content in the speckle; the ideal condition of equal numerical aperture, also for both lenses, is not usually adopted because of the inconveniently long focal length of L_R under such conditions;

(iv)   an optical wedge is used on the front of the camera tube to avoid the optically flat tube face from acting as a Fabry-Perot interferometer;

(v)   when operational, the reference beam and spatial filter are adjusted to give a uniform field at the camera, this adjustment being made whilst viewing the monitor; the speckly object beam is then switched in and adjustments made to the reference beam for optimum speckle contrast;

(vi)   the aperture of the camera lens L_C is adjusted to the smallest 'f' number possible consistent with the speckle pattern being resolved. Adjustment of the electronic video filter is then made to a cut off frequency corresponding to the smallest resolved speckle and a bandpass of about 0·8 MHz. The system is then ready to use.

As a guide, the system in the Loughborough laboratory employing a KGM camera uses a 100 mm focal length camera lens, L_C operating at f 30. The camera is fitted with a standard vidicon tube. Higher resolution camera equipment enables use of a smaller 'f' number, resulting in clearer fringe detail and more economic use of light.

If an interferometer is set up according to Fig. 9(b), the same principles should be adapted as far as possible. The need to pass both the object beam and the reference beam through holes in the mirrors will often necessitate compromise in setting up initially to minimise spatial frequency.

The setting up procedure for the interferometer is the same for all applications, but uses other than time-averaged vibration studies are less affected by the optical frequency content of the speckle.

8.2. Vibration of Large or Small Amplitude Compared to λ

The possibility of increasing the sensitivity of the interferometer has already been mentioned and is accomplished by introducing a modulator between the beam splitter and lens L_R in the reference beam path. The modulator can be driven at the same frequency as the object vibration, but viewing conditions are improved if a beat frequency of 3 or 4 Hz is introduced. At the other extreme the vibration amplitude may be too large to generate a visible fringe pattern. In this case information may still be obtained by introducing the electro-optic modulator in the illumination beam before lens L_B. The action here is to consider the fringes generated by just a small part of the motion (Fig. 10).

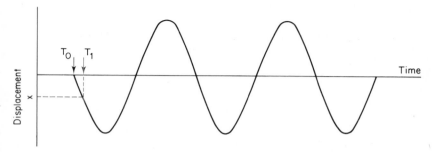

Fig. 10. Effect of pulse width modulation. A typical gating pulse would be at T_0 to open and T_1 to close in which time the maximum component displacement is x, where x may be between zero and the full vibration amplitude depending upon the time interval (T_1-T_0).

This use of the modulator also has the effect of introducing a reference for the vibration phase enabling components of coupled vibrations to be seen.

The technique was initially introduced in the author's laboratory to enable large amplitude vibrations to be examined for their mode structure. Variation of the switching period is used to adjust the sensitivity. This so-called pulse width modulation also yields fringes of good quality since the integration leading to the J_0^2 intensity pattern no longer applies. A typical result for a square plate vibration viewed by both methods is shown in Fig. 11. When using the pulse width modulation technique, nodal positions can be obtained without ambiguity by turning the mark space ratio of the modulator to unity, thus giving the time-averaged condition which is

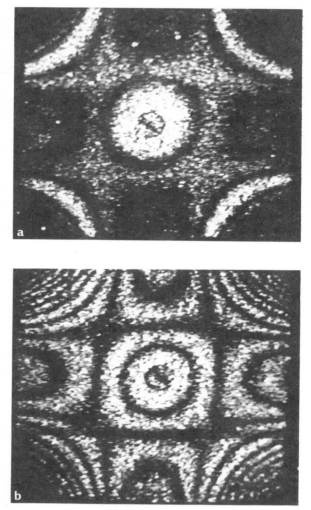

Fig. 11. Typical results from vibration mode viewing by ESPI (a) by the time-averaged process, (b) by the pulse width modulation process. The vibrating component in this case was a square steel plate excited by use of a piezoceramic crystal attached to the rear.

particularly favourable for node identification. A completely portable apparatus using the ESPI principles and suited to both time average and real time interferometry is made by Loughborough Consultants.

9. THE SPECKLE PATTERN COMPARATOR

In the early days of holographic interferometry after many enquiries from potential users it was required that a shape recording for one object could be compared with the shape of another similar object. This was not possible generally due to differing scattering properties, only specular or mirror like surfaces offering the possibility of direct interferometric comparison. Archbold *et al.*[12] demonstrated a comparator technique for internal combustion engine cylinder bore inspection utilising the fact that the honed surface would behave like a mirror if viewed at the glancing angle. The method, whilst interesting, was never adopted industrially, presumably because of its complex optical arrangement and limited applicability. At about the same time, results were being published[13] on the use of doubly exposed holograms, using different effective wavelengths at each exposure. The wavelengths λ_1 and λ_2 were either derived from two separate laser spectral lines or by changing the optical density of the media in contact with the object between exposures. In either case, setting λ_1 and λ_2 as the effective wavelengths, the resulting hologram, when processed and reconstructed, shows an image crossed by contours of interval d_c where:

$$d_c = \frac{\lambda_1 \lambda_2}{2[|\lambda_1 - \lambda_2|]}$$

(Note that for phase objects the magnitude of d_c is doubled for the same values of λ_1 and λ_2.)

These contour intervals form at equal intercepts along the line of sight and correspond to the beat frequency interval for the two waves recorded by the hologram. In principle this offers a means of shape comparison by simply examining the position relationship of the contours on each object to be examined. There are, however, some disadvantages. Firstly the contours form as planes (or spherical caps) in space, thus to examine an object of any considerable depth requires the use of contours of large interval d_c in order that they can be resolved in the image. This inevitably results in low measurement sensitivity, the difficulty being fundamental in that the possible measurement sensitivity is a function of object depth. An early publication of the technique illustrated the results by showing depth contours formed on an image of a silver dollar. The other disadvantage of the technique relates to the need to produce a hologram for every test, this being costly and time-consuming.

Speckle pattern interferometry offers a partial solution to this problem by enabling shape differences to be examined almost independently of object depth. The sensitivity, d_c, of the contour fringes formed remains as for holography, the mathematical basis of operation being given by Butters *et al.*[14] There are three areas of operation worthy of note: application to flat surface testing, application to regular geometry such as the cylinder or sphere and the application to complex geometry such as an aerofoil. The interferometer is set up as shown in Fig. 9(b) and to examine a nominally flat surface a plane collimated illumination beam is used. The intensity, I, at any position on the vidicon is determined by the phase at the beam splitter ϕ_b, taken as a convenient reference, the respective amplitude and phase of the reference beam, A_R and ϕ_R, the amplitude and phase of the signal wave A_S and ϕ_S, together with the path differences, which are the distance of object from beam splitter (g), the distance from object to vidicon (h) and the distance of reference beam point of divergence to vidicon (j):

$$I = A_R^2 + A_S^2 + 2A_R A_S \cos\left[\phi_R - \phi_S + \phi_b + \frac{2\pi}{\lambda}(2g + h - j)\right]$$

The fringe pattern seen on the television monitor corresponds to the summation of many speckles and assuming an RMS brightness distribution from the combination of two speckle fields with point values I_1 for λ_1 and I_2 for λ_2:

$$I_{RMS} = [(I_1 - I_2)^2]^{1/2} = 2A_R A_S \left|\cos\pi\left(\frac{1}{\lambda_1} - \frac{1}{\lambda_2}\right)(2g + \phi_x)\right|$$

where ϕ_x takes care of the resultant phase due to constant qualities in the test, i.e. ϕ_R, ϕ_0, h, j.

I_{RMS} then varies as $2g$, the periodic variation being given by

$$g = \frac{\lambda_1 \lambda_2}{2|(\lambda_1 - \lambda_2)|}$$

which defines the contour interval previously described as d_c.

The technique is particularly applicable to the rapid testing of flat ground surfaces where any damage is rapidly located and quantified.

Application in principle to other surface forms is the same except that the plane illuminating beam is replaced by a wavefront of nominally the same shape as the object under investigation. For a spherical surface wavefront shaping can be accomplished with a lens and for cylindrical surfaces a collimated beam set axially with an annular conical mirror of 45° cone angle will enable viewing of the cylindrical surface when its axis is also common with the mirror axis. For the third case cited, it would not be practical to generate a wavefront, using either mirrors or lenses, to match the surface of an aerofoil. In this case the object is illuminated by a wavefront generated holographically. As in the previous two cases the surface illumination should be incident in a well-defined direction preferably utilising the specular direction of illumination as close to surface normal viewing as possible. When using the holographic system a hologram is first prepared from a component called the master. This master may be a precision component, but it is equally valid to use any component from the batch, it only being necessary to obtain the shape data of this component to within the specified measuring tolerance using classical metrology methods or more conveniently a coordinate measuring machine. The master component must be given a specular or mirror-like surface before use to form the appropriate holographic optical element, any scattered light at this stage being faithfully recorded by the hologram and reconstructed later to result in speckle decorrelation and loss of fringe contrast. The surface preparation will depend on the initial condition, only light polishing being required, but in most cases performance is improved by lightly coating the surface with aluminium by vacuum deposition. With rougher surfaces showing pronounced crystal structure, the surface should be filled with resin before aluminium coating. One may be suspicious of the latter process for metrology, but checks in the laboratory have shown that when carefully treated with thin epoxy resin, the layer thickness is always below $3\,\mu m$ and the surface aluminises to form an excellent specular reflector. In the more probable case of using any component as the basis of comparison, the conventional metrology process should be applied after the hologram has been made since this ensures that the measurement relates to the coated surface and any scratches in the aluminium caused by the stylus will not cause extra scatter for the holographic recording.

When the master component is prepared, holograms are recorded for both wavelengths λ_1 and λ_2. It is better to record these holograms superimposed than to record them separately because errors due to lack of flatness of the photographic plate are cancelled in this way. The superimposed holograms are recorded with widely separate reference beam angles and in order that the real image may be subsequently reconstructed from the holograms without magnification error the reference beam used in recording is accurately collimated. The need to provide reference beams for recording and reconstruction beams subsequently introduces additional complexity in the interferometer but from a user viewpoint, measurements are easily made with the instrument, it only being necessary to place components in position to be viewed by the interferometer. With a potential repetition rate of 10 measurements per second, the instrument offers excellent possibilities for routine comparator application.

10. SPECKLE PATTERN TECHNIQUES FOR SURFACE FINISH MEASUREMENT

The form of light scattering from an optically rough surface is affected by the roughness form. In the case of random roughness the speckle contrast may be used as a measurement of this roughness on an arbitrary scale. Where regular marks exist, for example, on turned surfaces, the speckle pattern is modulated by a regular diffraction pattern. In order to establish a surface finish measuring technique, it is necessary to define a method showing a high degree of repeatability, at which stage calibration would be possible.

Optical physicists have been most active in studying the scatter properties of roughened surfaces. The main interest has been the scatter phenomena and for convenience, most experiments have been on roughened glass using the transmitted or forward scattered beam. The majority of actual surface roughness measurements are however conducted on opaque materials necessitating the use of the back scattered light. A thorough review of techniques for estimating the statistics of surface roughness has been conducted by Welford[15] who has considered the relations between known surface statistics and the near and far field scattered light statistics. Both monochromatic and polychromatic illumination methods are employed, but the complexity of the statistical process inevitably shows difficulties in forming any theoretical relationship with classical stylus methods. Wykes[16] has considered the case for measuring the change in speckle contrast with wavelength change on a theoretical basis. A proposed model fits well with experimental results. Cookson (unpublished) at Loughborough carried out a series of measurements on specially prepared surfaces with an ESPI instrument by integrating the net signal left when speckle patterns from the same surface at two different wavelengths were subtracted. Digital read-out was used and the method showed some promise for repeatability. The high resolution of the instrument could not be fully used, however, due to electronic noise, but further work is planned. The main difficulty with polychromatic methods lies in the need to use lasers of variable spectral output which place a high cost penalty on any resulting instrument. Single wavelength methods however suffer from severe range restriction running into trouble when the surface roughness feature size approaches the illuminating wavelength.

As yet there seems to be no method available that can be incorporated into a general purpose instrument, but experimental results have been obtained with high repeatability and showing reasonable correlation with stylus methods. The advantages to be gained for machining processes, by using non-contacting surface finish monitoring, will justify continued research on the topic.

11. NOTES ON APPLICATIONS OF SPECKLE

Perhaps the most refined measurement technique using speckle makes use of the speckle pattern comparator. In this application speckle methods perform a function which cannot be directly performed with any other interferometer. Once set up the apparatus could be used with photographic recording of the speckle interferograms, but the development has been carried out with the electronic speckle pattern interferometer which is almost essential for the initial setting of the optical system. The main value of the instrument lies in its ability to check many samples from a production run and the rapid read out from ESPI is invaluable here. An alternative requirement for checking long term shape stability, for example in plastic mouldings, is also met by this instrument which may be considered to have a memory in the form of holographic masters. Present work is seeking to link the ESPI directly with a computing facility to quantify the present analogue output. The resolution of this instrument can range from approximately $2\,\mu m$ to several millimetres, although a finely ground component surface is necessary for the higher sensitivity. The actual sensitivity intervals are determined by the available laser wavelengths, the argon laser having been employed mainly for this work. If only low sensitivity comparisons are required, developments of a moiré method would be more cost effective. The most attractive working range of the speckle pattern comparator is in the 5–$100\,\mu m$ tolerance bands.

Speckle methods are applicable in the more general aspects of non-destructive testing. Speckle photography is applicable to the testing of building and bridge structures,[8,9] although results are somewhat qualitative. Gregory[10] has used the method to test large pressure vessels claiming quantitative analysis. There have also been numerous applications to smaller civil engineering sub-assemblies, large castings, motor vehicles and structural steel work, to mention only representative examples. The method is attractive for its simplicity most commonly applied using a fairly standard, although good quality, 35 mm camera and a small He–Ne laser. Work by Burch at NPL even dispensed with the laser as previously mentioned by using surface marking. The subsequent processing of the photographic negatives requires some optical laboratory capability, but standard equipment for this work is now available.

On smaller structures higher resolution and a greater degree of repeatability are often called for. Speckle photography can still be applied, but the simple use of a camera may not produce adequate repeatability or resolution. The technique of speckle pattern interferometry should then be considered. When this stage is reached controlled conditions are required and parallel consideration of holographic methods is advised for higher fringe visibility. Speckle pattern interferometry is likely to be the best choice if it is desired to determine displacements along a particular direction or if a television camera is to be used instead of photographic recording. Speckle interferometry is used for the general NDT of plastic components, for the testing of bonds particularly in laminates and for vibration mode determination. It is also applicable to flow and convection studies. Although there has been interest in using the method for crack detection in metals, the reliability of application techniques to date has not been adequate, although there is promise of future developments. Sensitivity ranges for speckle pattern interferometry are the same as for holographic interferometry nominally $0.25\,\mu\mathrm{m}$ of movement between fringes. With speckle photography quantification is not generally possible, but movements from a few microns to several millimetres can be detected and sometimes measured.

Speckle pattern methods can be used to measure elastic strain. The big advantage over, say, strain gauges is that no surface contact is required and also the strain field can be displayed. Speckle photography will yield results with convenient apparatus and techniques, but special precautions will be required for accuracy better than 1000 microstrain even with bench size components. Speckle pattern interferometry techniques will do much better than this but are limited at the higher range due to decorrelation effects. The ESPI technique allows the dynamic display of strain fields but needs a stepped range, each step covering approximately 100–1000 microstrain. A comparison of interferometric and speckle pattern methods for measuring strains has been made by Dändliker[17] and Chiang *et al.*[18] have used white light speckle patterns for measuring strains around holes in cylinders.

REFERENCES

1. Lord Rayleigh. *The Scientific Papers of Lord Rayleigh*, Vol. 1, 1964, Dover Publications, New York, pp. 491–6.
2. Leendertz, J. A. Interferometric displacement measurement on scattering surfaces utilising speckle effect, *J. Phys. E.*, **3** (1970), 214, 218.
3. Goodman, J. W. Statistical properties of laser speckle. In: *Topics in Appl. Phys.*, Vol. 9, ed. J. C. Dainty, 1975, Springer-Verlag, pp. 9–74.
4. Ennos, A. E. Statistical properties of laser speckle. In: *Topics in Appl. Phys.*, Vol. 9, ed. J. C. Dainty, 1975, Springer-Verlag, Berlin, p. 208.
5. Jones, R. The design and application of a speckle pattern interferometer for total plane strain field measurement, *Optics and Laser Technology*, October 1976, 215–19.
6. Archbold, E., Ennos, A. E. and Virdee, M. S. Comp. paper: Speckle photography for strain measurements—a critical assessment, *SPIE J.*
7. Cookson, J. C. and Koyuncu, B. *Semi-automatic measurements of small high frequency vibrations using time-averaged electronic speckle pattern interferometry*, Internal Report, Loughborough University of Technology, publication to follow.
8. Burch, J. M. Large scale deformation measurement by a photographic moiré technique, *Nelex '74*, *Metrology Conference*, October 1974, paper 22.
9. Burch, J. M. and Forno, C. A high sensitivity moiré grid technique for studying deformation of large objects, *Optical Engineering*, April 1975.
10. Gregory, D. A. In: *Speckle Metrology*, ed. R. K. Erf, 1978, Academic Press, New York, Chapter 8, pp. 183–222.
11. Stetson, K. A. Problems of defocusing in speckle photography, its connection to hologram interferometry and its solutions, *J. Opt. Soc. Am.*, **66** (1976), 1267–71.
12. Archbold, E., Burch, J. M. and Ennos, A. E. The application of holography to the comparison of cylinder bores, *J. Phys. E.*, *Scientific Instruments*, **4** (1981), 489–94.
13. Haines, K. A. and Hildebrand, B. P. Multiple wavelength and multiple source holography applied to contour generation, *J. Opt. Soc. Am.*, **57** (1967), 155–62.
14. Butters, J. N., Jones, R. and Wykes, C. In: *Speckle Metrology*, ed. R. K. Erf, 1978, Academic Press, New York, pp. 146, 155.
15. Welford, W. T. Optical estimation of statistics of surface roughness from light scattering measurements, *Optical and Quantum Electronics*, **9** (1977), 269–87.

16. Wykes, C. Decorrelation effects in speckle pattern interferometry: I: Wavelength change dependent decorrelation with application to contouring and surface roughness measurement, *Optica Acta*, **5** (1977), 517–32.

17. Dändliker, R. Holographic interferometry and speckle photography for strain measurement: a comparison, *Optics and Lasers in Eng.*, **1** (1980), 3–20.

18. Chiang, F. P., Liu, B. C. and Lin, S. T. Multi-aperture white light speckle method applied to the strain analysis of cylinders with holes under compression, *Optics and Lasers in Eng.*, **2** (1981), 151–60.

Chapter 8

LASER ANEMOMETRY

R. G. W. Brown and E. R. Pike

Royal Signals and Radar Establishment, Malvern, UK

1. INTRODUCTION

Laser anemometry is used for the precise measurement of the movement of matter in its various states, solid, liquid and gas. It is therefore of interest to many people and has become widely practised.[1] Velocity, turbulence intensity, shear stress in fluids and many other quantities may be measured, yet in no way disturbing the motion of interest or having to employ calibration, as with mechanical probe techniques.

The range of studies now using laser anemometry is vast, from measurements inside internal combustion engines to measurements inside retinal blood vessels, from studies inside supersonic wind tunnels and jet engines to plant growth. The range of velocities involved is equally large, from microns per minute to very high Mach numbers, all measurable with essentially the same equipment.

Much of the work published to date has concerned laser anemometer theory, optical and signal processor designs and comparisons. It has come from research establishments worldwide. Recently more papers have appeared concerning specific applications.

It is our intention in this chapter to introduce the basic principles and problems of laser anemometry from a practical viewpoint, detailed theory being adequately covered elsewhere.

LASER DOPPLER ANEMOMETRY

2.1. Optical Systems

The most commonly used type of laser anemometer is the laser Doppler anemometer or velocimeter (LDA or LDV), employing optics known as the 'Doppler-difference' or 'real fringe' arrangement. A typical layout is shown in Fig. 1.

In this method a continuous laser beam (usually monochromatic, plane polarised and in the TEMoo mode) is divided by a beam splitter into two beams of equal power which are made to converge and cross at their waists. The waist position is determined by the lens L_1.

The region formed by the intersection of the two laser beams may be called the *probe volume*. The laser, lens L_1 and beamsplitter are mounted such that the probe volume is arranged to lie inside the flow at the point of interest.

The flow must contain small particles of matter (seed) that will faithfully follow the flow essentially instantly accelerating and decelerating with it. These particles will scatter laser light in a lobed pattern determined by their shape from both beams. Light scattered from the two beams will have a very slightly different Doppler shift of frequency. This is due to the slightly different angle made by each beam to the direction of motion of the particles.

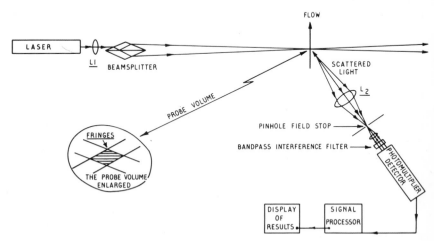

Fig. 1.   Laser Doppler-difference or real fringe anemometer.

The scattered light is collected by a good quality lens, L_2 (Fig. 1) and eventually reaches a detector, where the quantity of interest is the difference of the frequencies of the light from the two beams, i.e. the Doppler frequency difference. In a number of earlier studies it was simply assumed that two waves were mixed on the photocathode which behaved as an analogue square-law detector. In such a case if the waves are given by $E_0 \cos \omega_1 t$ and $E_0 \cos \omega_2 t$, taking the amplitudes to be equal for simplicity, the output of the detector would be

$$I(t) = E_0 (\cos \omega_1 t + \cos \omega_2 t)^2$$

$$= \frac{E_0}{2} [2 + \cos 2\omega_1 t + \cos 2\omega_2 t + 2(\cos(\omega_1 + \omega_2)t + \cos(\omega_1 - \omega_2)t)]$$

and analogue filters would pass only the difference-frequency term. More detailed considerations which take proper account of the quantum theory of optical detection[2] show that, for the above case, the output of the detector is, in fact, a Poisson point process, rate modulated by the function

$$I(t) = 2(1 + \cos(\omega_1 |- \omega_2)t)$$

The sum- and double-frequency terms are thus never present and the detector output is a digital trair of pulses corresponding to detections of single quanta of light. A further time dependence is given by the fact that the plane waves are not of spatially infinite extent but have a Gaussian intensity cross-section. Figure 2 shows the modulating function and the corresponding random train of pulses which constitute the real output of the detector. Any finite response time of the detector or its following

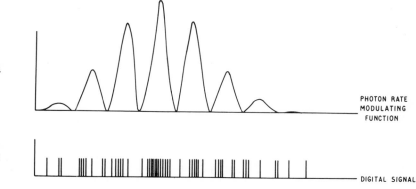

Fig. 2. Signal due to a scattering particle.

circuitry will broaden the pulses and if they are allowed to overlap in time, quasi analogue d.c. shifts will be built up. In the limit of a very large photon rate this can give rise to a quasi-continuous analogue signal with the approximate form of the modulating function itself. We shall discuss later methods of signal processing which utilise the direct photon train output of a good quality high-speed photomultiplier (photon-correlation processing) and methods which are based on the quasi analogue signals which can be obtained at high light levels (analogue processing).

An alternative and popular description of the Doppler-difference frequency is called the real-fringe model. It is widely used for day to day anemometry calculations. If one greatly magnifies the probe volume then Young's interference fringes may be seen on a screen (Fig. 3). These fringes do not exist in the probe volume as no non-linear optical processes are occurring there; the squaring to produce the interference pattern occurs only in the eye. Assuming that fringes are present, however, then one imagines that as the particles pass through the probe volume larger numbers of photons are scattered from the bright fringes and almost no photons from the dark fringes; the frequency will be given by the rate at which the particle crosses the fringes and this comes out to the same formula as the Doppler-difference calculation. Consideration of a particle

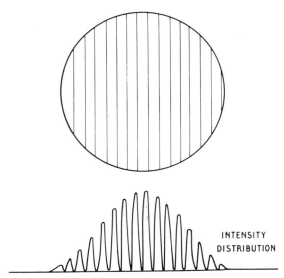

INTENSITY
DISTRIBUTION

Fig. 3.   The probe volume cross-section filled with fringes.

with a null scattering between lobes at one of the Doppler angles will show the fallacy of the model.

Using the fringe model (with caution) for convenience, we may imagine a perfectly steady flow (constant velocity, zero turbulence) passing through the probe volume, and the detector's analogue output in the high photon-rate limit due to a single particle is as shown in Fig. 2(a), namely, a single frequency oscillation with a Gaussian envelope. If, however, the flow varies in speed or direction (turbulence) then so will the frequency of the detector output. This variation in frequency is proportional to the variation in the velocity of the flow.

It must be stressed how important it is to have a probe volume filled with fringes that are separated by exactly the same distance. Variations lead to errors in the estimation of mean velocity and turbulence. Such errors may occur for two reasons. Firstly, if the beams do not cross and waist at the same point then fringe divergence occurs (Fig. 4) causing a constant velocity to appear as a spread of velocities, depending on where scattering particles cross the probe volume. The lens L_1 in Fig. 1 is used to correct this condition by appropriate choice of focal length and position. (Further details are given in the Appendix.) Secondly, in flames for example, refractive index fluctuations may occur along the laser beam paths causing fluctuations of the probe volume size and fringe spacing.

The measured quantity in the fringe arrangement is the *component* of velocity perpendicular to the fringes (i.e. parallel to the plane in which the two laser beams lie). Using one set of fringes, one component of velocity is

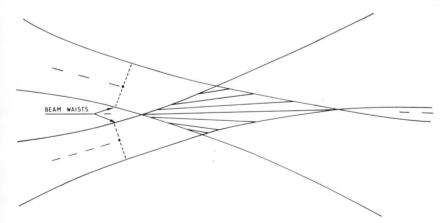

Fig. 4. Intersection error of two laser beam waists.

measured. Two- and three-component systems[3] may be constructed using a multiwavelength laser, two and three orthogonal sets of fringes being superimposed. The complexity of three-dimensional systems, however, has seriously restricted their use up to the present.

It is useful in fringe experiments to calculate three quantities. Referring to Fig. 5:

1. Beam waist diameter at the probe volume

$$d \simeq \frac{4}{\pi} \cdot \frac{\lambda F}{D} \tag{1}$$

2. The number of fringes between the e^{-2} points of the probe volume

$$N \simeq \frac{4}{\pi} \cdot \frac{l}{\delta} \tag{2}$$

3. The fringe spacing

$$S = \frac{\lambda}{2 \sin\left(\dfrac{\theta}{2}\right)} \qquad \left\{ \simeq \frac{\lambda Q}{P} = \frac{\lambda}{\theta} \text{ for small } \theta \right\} \tag{3}$$

where $F =$ focal length of the lens, $D =$ the beam diameter at the lens, $l =$ spacing between beams, whose diameters are δ, $\theta =$ the angle between the beams and $\lambda =$ the laser light wavelength.

One must not only be careful in creating the probe volume, but also be careful in handling the collected scattered light.

The collecting lens may be positioned virtually anywhere with respect to the probe volume, remembering, however, that little or no scattered radiation will be found in or near the plane of polarisation of the laser beams, and that the scattered power angular distribution depends heavily on the laser light wavelength and scattering particle size. The forward scatter power is often a thousand times more than the backscattered power.

The collection lens should be of such quality to allow good imaging of

Fig. 5.   Details of Doppler difference geometry.

the probe volume at the field stop. The field stop is usually a pinhole matched (not less than twice the $1/e^2$ points because of truncation errors) to the probe volume image size.

Between the field stop and detector the light may be collimated in order to pass through a narrow band interference filter (say 20 Å bandwidth and greater than 50 % transmission) which blocks all wavelengths to which the detector may respond, except the region of the laser light wavelength.

The collected light is finally made to fall onto the detector, whose spectral response should be a maximum for the wavelength of light being used. Photomultipliers employing S20 photocathodes are popular due to their good He–Ne laser wavelength performance; S11 are preferable with argon lasers. Further, the detector should have a low dark current (dark count) and, particularly when signal processing with a photon correlator, no spurious afterpulsing correlations or oscillations of its own (or of its high voltage supply).

The effects of spurious detector properties can be seen and eliminated when using photon-correlation processing; the magnitude of these effects on other types of processor are unknown since they are smaller than other analogue system errors.

The arrangement described above is a good general purpose optical system for working with all except high scattering particle density conditions and up to distances of say 100 m (using increased laser power). Alternative optical arrangements to achieve the Doppler configuration are described elsewhere.[4]

No mention has yet been made of the determination of flow direction. The Doppler signal will appear the same for flows of equal velocity and opposite directions. To solve this problem and to increase the high turbulence measurement capability of signal processors, various devices have been used to move the fringes in the probe volume in a known direction, providing a bias velocity. Two popular methods employ either Bragg (acousto-optic) cells[4] or a Pockels (electro-optic) cell.[5] The Bragg cells, at present, provide a greater shift of the Doppler frequency (up to 7·5 MHz) but require realignment of the optical system when the frequency shift is varied, as the probe volume is moved in space, unless a pair of cells in series is employed.[6] The Pockels cell method is easy to align and provides a probe volume fixed in space for all frequency shift values. Presently available Pockels cell devices, however, can only shift the Doppler frequency by up to 1 MHz which although sufficient for many users is not sufficient for very high speed, high-turbulence situations. When using either method of frequency shifting the devices are placed between the

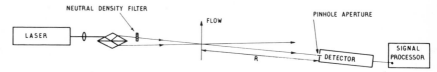

Fig. 6. A reference beam laser Doppler anemometer.

beamsplitter and probe volume (Fig. 1). It should be noted that when using a photon correlator to process the Doppler information it is possible, by slightly altering the optical arrangement, to find flow direction without the use of the frequency shift devices just described (p. 307 of ref. 7).

Two other Doppler configurations are useful. The Doppler difference system works (ideally) with scattering particle densities of up to one particle in the probe volume at a time (light seeding). However, situations may be encountered where two or many more particles will always be in the probe volume. Signals from these scatterers will interfere with each other and reduce the system performance. Under these circumstances, the reference-beam Doppler arrangement may be advantageous[4] (Fig. 6). Light scattered from one laser beam is mixed with a very weak amount of the original laser beam. No collecting lens need be used, in which case the field stop is chosen such that only light from the beam waist mixes coherently on the detector. The correct field stop area, A, is obtained from the equation

$$A = \frac{\lambda^2 R^2}{A_p} \tag{4}$$

where λ = wavelength of the laser light, R = distance from the focal point to the stop, A_p = cross-sectional area of the probe volume. A lens and field stop, identical to those used in the Doppler-difference method, placed in line with the probe volume and detector will complicate the system somewhat but will improve the signal to noise ratio.

The other useful optical arrangement for Doppler anemometry is the monostatic heterodyne (homodyne) method, which gives information about the velocity component along the laser beam direction, the line-of-sight velocity.[8] A typical backscatter arrangement is shown in Fig. 7. A single beam of laser light is emitted. Light scattered by particles moving along or obliquely across the beam is Doppler shifted and is mixed with a few percent of the original laser beam to extract the Doppler information. Using visible lasers, the Doppler shifts will exceed those capable of being handled by current signal processors when the in-line velocity is only a few metres per second. It is therefore usual to use this method with infrared

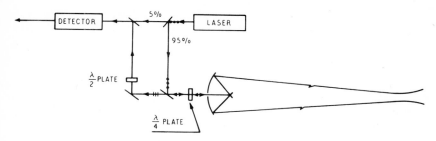

Fig. 7. Line-of-sight single beam heterodyne laser anemometer.

CO_2 lasers whose wavelength is $10 \cdot 6 \ \mu m$ thus lowering the Doppler shift for the same velocity.

The Doppler shift is given by

$$f = \frac{4\pi}{\lambda} \ v \cos \theta \qquad (5)$$

where θ is the angle between the flow of velocity v, and the line of sight. The high power and eye safety of CO_2 laser radiation make this type of anemometry attractive for long range work. Transverse velocity information may also be deduced,[9] but has not proved very practical.

2.2. Signal Processors

A variety of signal processors have been used to derive flow velocity and turbulence parameters from the Doppler information acquired through the use of the above optical systems. In alphabetical order, the methods are:

1. burst counting;
2. Fabry-Perot interferometry;
3. filterbank;
4. frequency (spectrum) analysis;
5. frequency tracking;
6. photon correlation.

A brief outline of these methods follows, with comments on their capabilities and error sources.

2.2.1. Burst counting
Burst counters require higher signal powers than other signal processors so that good bursts of quasi analogue Doppler information (Fig. 2(a)) occur

for each particle that traverses the probe volume. The signal is first low-pass filtered to remove as much of the beam transit time effect as is possible without distorting the signal excessively, and then the time between the zero crossings of the signal is estimated.

The electronics must be carefully constructed to reject as far as possible various errors. The zero crossings may be of unequal time intervals due to residual transit time effect or finite fringe number. Spurious zero crossings may occur due to noise, or zero crossings may be missing due to, perhaps, phase reversal during the signal if a second particle enters the probe volume. For these reasons, various *ad hoc* electronic tests are performed on the signal to reject apparently incorrect bursts, e.g. comparison of 5 cycles with 8 cycles of Doppler. Two electronic biases require consideration, (i) velocity–amplitude bias, i.e. the rejection of signals from higher velocity, lower signal amplitude particles because they fail to exceed a preset discrimination level and (ii) polar response bias, i.e. the rejection of signals from particles travelling nearly parallel to the fringes and intersecting too few for acceptance. The requirement of electronic tests and good signals makes burst counting potentially inefficient, often only operable in forward scatter and the analogue errors introduced are not easily quantified.

Unlike the frequency tracker to be discussed shortly, the burst counter does not require a continuous signal and can function with flows containing only a few scattering particles. In turbulent flow conditions the velocity results obtained need correction due to an effect known as velocity biasing. When the instantaneous velocity is greater than the true mean velocity, more particles will pass through the probe volume than if the velocity is constant, and vice versa. The measured velocity probability density will be biased towards a higher velocity.[10,11]

Currently available processors accept Doppler frequencies from 1 kHz to 200 MHz giving a wide velocity range depending strongly on the optical geometry. Mean velocity results may be correct to within a few percent error for good signal to noise ratios and moderate turbulence intensities. A more detailed description of burst counting is given in refs 4 and 12.

2.2.2. Fabry-Perot Interferometer

The Fabry-Perot interferometer (FPI) has been used on occasions to measure flows of high velocity.[4,13] Normally the scattered light and a reference beam are directed into a confocal FPI. One of the mirrors is piezoelectrically scanned with a sawtooth voltage, and the spectral output of the FPI is displayed on an oscilloscope.

Flow velocity and turbulence may be estimated but the precision of the method depends on the finesse of the interferometer and the stability of the laser; flows of a few hundred metres per second may be measured with errors of the order of a few percent.

The method uses collected light very efficiently but requires a highly stable single-mode laser and delicate optics. The method is overlapped by photon correlators and burst counters which can easily measure high velocities, are more rugged and can yield more and better quality information through fast and simple data processing.

2.2.3. The filterbank
This is a rarely used method of analysing the Doppler signal by passing it through a bank of fixed parallel bandpass filters which slightly overlap. The method, although efficient, is inflexible and the analogue nature of the store is restrictive and difficult to set up and maintain.

Further details if required may be found in ref. 4.

2.2.4. Frequency analysis
Spectrum analysers for general purpose analogue signals have been available for many years. Certain special requirements must be met for their use in laser Doppler anemometry. One requires an *a priori* knowledge of the spectrum to be measured in order to set a sufficiently large sweep width. Further, the width of the intermediate frequency filter should be small compared to the width of the spectrum to be measured.

When using a spectrum analyser one must arrange for a large number of fringes in the probe volume in order to reduce the errors due to transit time broadening of the spectrum. This may reduce the maximum measurable velocity. Further errors may occur due to the dependency of the Doppler spectrum on signal amplitude as well as frequency. Careful interpretation of the data is required to yield accurate results.[4,12,14] The method is normally a single channel one looking at frequencies serially and is thus very inefficient.

The spectrum analyser does not have the limitation of the burst counter and frequency tracker of requiring large signals to work at all. It will produce results by using long enough integration at all signal levels including digital photon processing. It must be remembered that the theoretical spectrum in this latter case lies on a white shot noise background component which will, however, in practice not be flat due to finite analogue response times and distortions. Experiments have been performed using surface acoustic wave spectrum[15] analysers which are efficient parallel processors and very suitable in particular for infrared anemometry where full digital photon processing is not possible due to the low photon energies in that region.

2.2.5. *Frequency tracking*

The usual method of frequency tracking involves mixing the quasi analogue Doppler signal with a sinusoidal voltage from a voltage controlled oscillator (VCO) which is in a feedback loop from the tracker output. The mixed signal is passed through a fixed narrowband filter and frequency discriminator to give an output voltage, which is used to drive the VCO. The result of this feedback arrangement is that the VCO frequency tracks that of the incoming Doppler signal, and the output voltage is proportional to the instantaneous Doppler frequency.[14]

The principal disadvantage of this processing method is that of signal dropout (loss of Doppler signal) which may be caused by low scattering particle concentrations, low signal amplitude or inability of the tracker to follow very fast Doppler fluctuations. Various methods may be employed to try to overcome such dropouts, particularly those of introducing large quantities of scattering particles into the flow, using higher powered lasers and shifting the Doppler frequency to a more convenient value for the equipment.

As with other frequency analysis methods, beam transit time broadening of the Doppler signal can also cause significant errors.[4]

A further source of error is phase noise, which occurs in the real fringe method when more than one particle contributes significantly to the Doppler signal (overseeding). The particles will have an arbitrary phase difference, causing the tracker output voltage to vary without relation to the flow velocity.

Trackers are often used to measure laboratory water flows where high scattering particle density can be arranged with low flow velocity and low turbulence frequencies; forward scattering is usually essential to obtain a high enough signal.

2.2.6. *Photon correlation*

The advantages of using photon correlation processing are that it acts on the pure digital photon signal with no analogue distortions and one can measure extremely accurately a very wide range of flow velocities, particularly in conditions of low signal intensity as is normally encountered in backscattering optical arrangements. Furthermore good velocity estimates may be obtained from about 1000–10 000 times less optical signal than is usual for other processors, enabling one to use very low power lasers, exceptionally low scattering particle density and, as a bonus, increasing eye safety. The technique does not suffer from velocity or large particle signal biasing (due to the sampling and processing arrangement) and the post-correlation data analysis can account exactly for beam transit time effects. A most important LDA consideration not widely appreciated is that whenever extra seed is added to a flow, errors in turbulence levels will be measured in any regions where mixing occurs with flows seeded at a different level.[16] The ability of the photon correlation method to measure in conditions of natural seeding often avoids this effect.

Perhaps the major problem that photon correlation has suffered during its years of use is the lack of a non-mathematical description of how a photon correlator works. The explanation is conceptually very simple.

The first major difference between correlation and the other forms of signal processing is in the output of the detector. Instead of an analogue output from the photomultiplier, the individual electron pulses are used (corresponding to detection of individual photons), meaning that the natural ultimate sensitivity is obtained. These pulses are passed through an amplifier and pulse shaper in order to make them suitable for transmission along cables and entry into the correlator.

The second major difference to be noted about the use of a photon correlator is that no conditioning and filtering of the input signal occurs, which are obligatory with most other techniques. The correlator operates by counting the incoming photon pulses in equal, adjacent, time intervals. No pulses are lost between sample time periods due to electronics known as derandomising circuitry. The value of the sample time used must be much less than the period of the Doppler frequency to be measured. Referring to Fig. 8, the samples of the pulse train are passed into a shift register in which they are clocked or stepped along by a sample time clock. After every step along the register, all the samples (now 'old' samples) are multiplied by the latest prompt sample value and the results are added to an integrating store. A new sample is now taken and the procedure repeated continuously at the clock rate.

Using the fringe model of the anemometry optics, one may imagine that bunches of photons (and thus photo-electron pulses) will be related to a particle passing through a bright fringe, with gaps between the bunches due to the dark fringes. The cosine wave depicted in the store (Fig. 8) is now explained as peaks being due to multiplications of samples of pulses due to bright fringes, and troughs due to multiplications of samples due to a bright and a dark fringe. The autocorrelation function of the incoming pulse train has been formed.

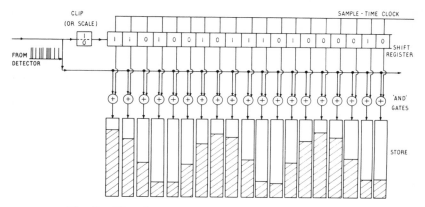

Fig. 8.   The photon correlator (autocorrelation mode).

There are many photoelectrons created in a typical experiment, perhaps 10 000 to 500 000 per second. In order to reduce the amount of data handling and storage (often not necessary with photon correlation anemometry) two techniques can be applied to the pulse train before it is correlated. One can either clip the signal (assuming Gaussian signal statistics) by setting a threshold number of counts to be reached each sample time before a '1' is sent into the shift register, otherwise a '0' is sent. Alternatively one may simply scale down the number of pulses to enter the shift register by dividing by a fixed number, using the photon statistics to randomise the clip level. This latter method is valid for all signal statistics and should be used for anemometry; but it is less efficient than clipping. It can be seen that considerably less manipulation and storage ability is required when clipping or scaling are employed. A rigorous description of photon correlation is given elsewhere.[2,7]

The correlation function built up by the sample and multiply process rarely consists of a single cosine wave. Flow turbulence causes a range of Doppler frequencies, and when these are processed they cause the cosine to decay. The rate of decay is a measure of the turbulence of the flow. The decaying cosine is superimposed on an exponential decay which is caused by the transit time of the scattering particles as they pass through the Gaussian intensity probe volume. A complete picture of a correlation function is shown in Fig. 9.

In order to extract the mean flow velocity and turbulence level, etc., post-correlation data processing is performed using an on-line desk top calculator or mini-computer. This extra step is warranted over simple

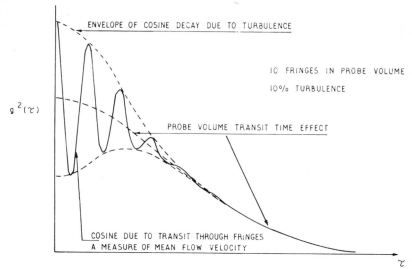

ENVELOPE OF COSINE DECAY DUE TO TURBULENCE

IO FRINGES IN PROBE VOLUME

10% TURBULENCE

$g^2(\tau)$

PROBE VOLUME TRANSIT TIME EFFECT

COSINE DUE TO TRANSIT THROUGH FRINGES
A MEASURE OF MEAN FLOW VELOCITY

τ

Fig. 9. A typical correlation function.

inspection of the frequency and damping since computer processing uses the intrinsic accuracy of the data to best effect. A review of data processing methods may be found elsewhere.[17,18]

Photon correlators allowing a minimum sampling time of 50 ns have long been available commercially. However a 10 ns correlator (and associated detectors) is available which can process in real time data rates up to 100 MHz and easily handle the Doppler frequencies from very high speed flows. Post-correlation processing may be performed either by the correlator's microprocessor controller or electronically to give mean velocity estimates at up to 20 kHz rate.[19]

It is difficult to test the accuracy of photon correlation measurements. Calculation of the parameter values required for a desired accuracy may be made by computer simulation[17,18]—experiments with <1% error are regularly performed—in many cases the limit is not due to the signal processing but to the precision with which the fringe spacing and probe volume diameter can be measured.

3. TIME-OF-FLIGHT ANEMOMETRY

Laser Doppler anemometry is well-established and widely practised yet the method, though extremely useful, has one state of limited operational capability. In many experimental configurations one can place the transmitter and receiver at almost any desired position. Consequently when attempting to make measurements normal and very close to boundary surfaces, such as solid walls, the receiver optics can be positioned virtually to eliminate the stray light reflections from these surfaces swamping the Doppler signal. However, when attempting measurements inside wind tunnels, turbo-compressors, turbines and internal combustion engines for example, mechanical constraints sometimes force one to place the receiver in the full or partial backscatter position, adjacent and almost parallel to the transmitter optics. Under these circumstances, to obtain results close (within say 5 mm) to any boundary is impossible as the receiver is filled with stray light.

One way in which this problem can be partially overcome is by reducing the size of the probe volume and the pinhole stop in the receiver. Unfortunately, with current signal processors and the requirement of high velocity information, there are minimum limits to the probe volume size. The smallest usable volume may be of approximately 100 μm diameter.

In 1968 an alternative technique to the Doppler system was proposed[20] and it has been developed and analysed[18,21-25] with the aim of conquering the stray light problem. The principle of this technique is to measure the time of flight (transit time) of scattering particles from one highly focused laser beam to another beam, parallel to and a little distance away from it, as shown in Fig. 10. The beam waist diameters are typically about 10 or 15 μm, and therefore the power density at the waist may be a few hundred times that of a laser Doppler arrangement (using the same laser power).

Because of the great reduction in the probe volume and the fast

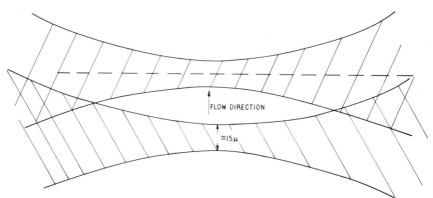

Fig. 10. The time-of-flight anemometer probe volume.

divergence of the beams beyond it, this technique may yield the required flare rejection (due to better spatial filtering) whilst allowing measurement of mean velocity with good directional accuracy. The price paid for the improvement in flare rejection is that far fewer scattering particles cross the probe volume especially in turbulent flow.[18] Furthermore, the probe volume must be carefully rotated to find the flow direction before measurement, and then a series of correlograms are needed at different angles of the probe volume to the flow direction in order to acquire sufficient information to yield reliable estimates. This procedure can be time-consuming in an unknown flow. It is important to be able to rotate the pair of waists to any desired flow angle. If two 15 μm waists spaced at say 500 μm are misaligned from a laminar flow by more than $\simeq 2°$, then no scattering particles can pass through both waists and no correlation peak is formed.

The authors have made, at 0·5 m range, medium velocity compressor measurements to within 1 mm of normal incidence, polished metal surfaces and to within 100 μm of dirty glass windows, also at normal incidence. The signal processor was a photon correlator; the two spot optical system is described in detail elsewhere.[24] The performance of both transit and Doppler anemometry may be improved by using fluorescent seeding and spectral filtering.[24]

It must be stressed that the shortest distance to a solid target at which measurements may be made depends heavily on the nature of the surface and the extent of the probe volume along the laser beam path viewed by the receiver. The type of material, its surface roughness, degree of reflection specularity, surface finish (matt, polished, anodised) and precise angle of incidence to the laser beams all significantly affect the ability to make velocity measurements. 3M Corp. 'Black Velvet' coated or matt black anodised metal surfaces are preferable. A further requirement is that sufficient scattering particles can be entrained into the boundary region being measured. The extent of the probe volume depends on the receiver optics quality (diffraction limited preferred), collection solid angle and the effectiveness of the spatial filtering.

Although the time-of-flight (often called two spot, laser two focus, L2F or transit) method allows mean velocity and direction measurement, it seems to provide a poorer means of estimating turbulence than LDA and rapidly becomes unusable above about 20 %. In flows of any significant degree of turbulence, cross-correlation of the signals from two photomultipliers is required; each photomultiplier receives light from a different beam waist.

When attempting measurements between rotating machinery blades, the signal processor is usually gated ON or OFF in synchronism with the blade passage time. This procedure avoids the very high flare due to blade passage through the probe volume and defines a precise, small region between the blades to which the measured velocity can be related. The ON time should obviously be longer than the transit time of the scatterers. Also when measuring inside machinery the probe volume may be deflected with respect to the original alignment of the transmission optics, due to the passage of the beams through a glass window; an internal reference position should be used.

Two rather different types of signal processor may be used for transit anemometry, the photon correlator operated in cross-correlation mode or a multichannel analyser (which, in fact, is also to be found as a standard mode of operation in commercially available photon correlators, though shorter in record length).

Cross-correlation of the signals from the two laser beam waists either at the photon level or by obtaining a pulse from the quasi analogue signal using a constant fraction discriminator is found to be a good method of data retrieval.[25] The use of a multichannel analyser in a non-gated situation is less efficient in situations of high seeding. In both cases it is required that scattering particles pass through *both* beam waists. Those particles that pass through only one beam (either beam) contribute no signal, only background noise. The detectors must, of course be connected in the correct order to the processor, to see results in positive time.

If multichannel analysis is used, then a sweep is initiated due to a particle passing through one beam waist and events occurring in the other beam waist are recorded in the appropriate time delay storage channel. In high turbulence situations many sweeps will be initiated by particles that do not pass through the second waist. During the now useless sweep time, other particles will pass through both waists, only to be recorded as increase in background noise. In cross-correlation, the distribution of probable event times of *all* particles that pass through the first waist and *all* particles that pass through the second waist is recorded. Cross-correlation is therefore more efficient in measuring high turbulence levels. In cross-correlation mode a factor of four improvement in flare rejection (rate of background level increase) is found, compared with autocorrelation of the signal from one photomultiplier that receives light from both laser beam waists. Furthermore, cross-correlation does not exhibit the baseline distortion that can be observed with multichannel analysis under conditions of high noise.[18]

A typical time-of-flight correlation or multichannel analyser function is shown in Fig. 11. A group of such functions acquired at various angles to the flow may be analysed by an on-line minicomputer or desk top calculator to derive mean velocity shear stress, flow angle and turbulence intensity, etc.

It has been mentioned in the description of laser Doppler anemometry that errors may be introduced due to refractive index fluctuations along the laser beam path such as in flames or through dirty optical windows. Such fluctuations destroy the wavefronts of the beams that are essential to form good fringes and may make Doppler measurements impossible. The time-of-flight anemometer is less susceptible to such imperfections, but they do cause broadening of the waists and give mean velocities, etc. with greater error. Optical access windows should be cleaned at every available opportunity. Methanol and Decon 90 are useful.

The time-of-flight anemometer is not affected by large particles, assuming that they faithfully follow the flow, whereas in the Doppler-difference anemometer such larger particles (diameter greater than fringe spacing) may cause signal degradation.

A final, significant source of error may occur if the receiving system is slightly misfocused and the two beam waists are not parallel to say better than 1°. Misfocusing will cause a broadening of the peak and lead to error in the turbulence estimation; the simultaneous lack of parallelism causes a systematic error of a few percent in the mean velocity.

In order to achieve perfect focusing one must locate the beam waists with a very thin ground glass screen or tracing paper, looking for the maximum

Fig. 11. The time-of-flight correlation function.

size of transmitted or reflected speckle.[27] The target should now be spun rapidly to average out the speckle. The receiver may now be correctly aligned without errors due to the coherent speckle imaging that will be found when using a stationary target. This technique is also useful in Doppler anemometry, but here the focusing is usually less critical.

Doppler and time-of-flight systems have been developed quite separately, each having their own advantages. However, an instrument has been constructed which allows both modes of operation using almost identical optics and allowing rapid interchange between the modes.[24]

4. SEEDING

All laser anemometry systems rely entirely on small particles (seed) to scatter light from the probe volume. It is the velocity of the seed that is measured and not necessarily the velocity of the medium in which the seed is suspended, unless care is taken to ensure that the seed is capable of instantaneously accelerating and decelerating with the flow velocity variations.

In Doppler-difference anemometry, the seed also must not be larger than a fraction of one fringe spacing, otherwise signal degradation occurs. Some LDV arrangements have been used to yield particle size information simultaneously with velocity measurement.[4] Such claims should be treated cautiously as the fringe model is used to explain the effect, yet the signal amplitude varies in a highly irregular fashion with size, shape and refractive index. This effect is not present in the reference beam system where full visibility is obtained at all particle sizes.

Besides faithfully following the flow, seeds must possess many other properties, being good scatterers of light, easily generated, non-toxic, non-abrasive and non-corrosive, as non-volatile as possible and if used in quantity in confined spaces, non-odorous and biodegradable for health reasons (vegetable oil is often used in place of silicon oil, where applicable). Ventilation (without disturbing slow air flows) is usually desirable.

Some straightforward equations may be applied to check that one's seed is suitable for a particular application (p. 251 *et seq.* of ref. 7), neither too large to lag, nor too small to be affected by Brownian motion. A further consideration is that of coagulation. It is difficult to predict coagulation and a check should be made periodically on the performance of one's seeding equipment. To avoid coagulation, it is useful to be able to seed with as low a particle density as possible, a point in favour of burst counters and photon correlators. A similar requirement may be necessary on health grounds as seeding usually used is completely inhalable.

For supersonic work particles of 1 μm diameter may easily lag behind the flow by some 15% or more; particles of about 0·3 μm diameter are often used, in spite of their lower scattering properties. Well below sonic velocity, particles of about 0·5 μm diameter are often acceptable for use in air and many commercial seeding equipments emit particles of about this size. Larger particles can easily be removed in a spiral separator of small radius. For combustion research a useful design of solid particle generator exists.[28] In liquid flows artificial seeding is often not required, but with very clear liquids polystyrene spheres are useful as an additive. A wide variety of methods of generating particles is described elsewhere.[4] These methods include atomisation, condensation, combustion and fluidisation. Silicon and vegetable oils are popular for atomisation and titanium dioxide or aluminium oxide solid particles for fluidisation.

5. SOME PRACTICAL CONSIDERATIONS

Engineering environments tend to be, in the optical sense, very dirty. For this reason optical components used in laser anemometers may need frequent cleansing with, for example, spectroscopic grade acetone or ethanol. Where dielectric mirrors or antireflection (AR) coatings are employed, they should be specified as 'hard' so that frequent cleansing will not degrade their performance.

In certain engineering situations, for example adjacent to aircraft engines, very large vibrations and noise will be transmitted to the optical equipment unless it is mounted on antivibration mounts and acoustically protected. These vibrations can shatter laser tubes and photomultipliers, cause serious misalignment of the optics and possibly be measured as spurious Doppler velocities. In the open air, environmental protection for the laser anemometer may be necessary.

Optics should not be placed too close to flows of hot gases or solids for fear of being cracked.

The size of laser required for anemometry depends heavily on the type of signal processor and the amount of seeding (scattering particles) that can be used. The least amount of seed is required not simply for health reasons, etc., but because seed particles can adversely affect machinery and make opaque optical access ports into machinery. Also, as mentioned above, erroneous turbulence levels will be obtained in mixing regions where a flow of different seeding density is entrained.

Laser powers might typically be a few milliwatts for lightly seeded flows of up to say $100\,ms^{-1}$ when using correlation for signal processing; somewhat higher powers or greatly increased seeding are required when using alternative techniques. Greater velocities may require greater laser power.

Plane polarised laser power can be conserved by ensuring that the lowest loss optics is employed. For example, the style of beamsplitter used in Fig. 1 has been perfected for Brewster angle input (p-polarisation) with a total power loss of less than 2 %. It also has the useful properties of a wide range of output beam separations and convergence angles, and no external, damageable AR coatings. Laser beams can be deflected by employing small prisms whose input and output angles are the Brewsters angle. These obviate the need for expensive and damageable dielectric mirrors.

Beamsplitters should be carefully constructed; the two output beams must eventually cross exactly, and therefore the divergence between the beams should be less than (typically) 10 seconds of arc.

To reduce the probe volume size for finer sampling of the flow one may replace the lens L_1 with a beam expander (Fig. 1). The demands on the beamsplitter are now increased.

Using signal processing techniques other than photon correlation and burst counting one must use perhaps 20 or more fringes in the probe volume in order to reduce the beam transit effects. However with photon correlation one may use from 3 fringes upwards at the expense of turbulence accuracy. This can often be a useful advantage. The fringe spacing can be increased to allow measurements of higher velocities, keeping the Doppler frequency within the range of the signal processor.

The interference filters used in the receiver optics are narrow band pass. Even hermetically sealed modern filters can 'age', the peak wavelength of transmission shifts, due to temperature and moisture change leading to greater power loss. It is advisable occasionally to check the transmission of the filters employed (the transmission peak also changes with angle of incidence).

The light incident on the photocathode normally has only one chance of being detected. However, by careful design one can insert either wedge prisms (large photocathodes only) to cause multiple reflections from the photocathode material using total internal reflection or, with small photocathode tubes one may employ multiple element and polarisation rotating devices for multiple reflections from the photocathode. Both methods can greatly increase the effective quantum efficiency of the photocathodes.

Finally, when transmitting Doppler signals or pulses over many metres to the signal processor one must be especially careful in a machinery environment about earthing and transients causing spurious signals.

6. APPLICATIONS

Laser anemometry may be useful wherever there is a need to measure velocity, fluctuations in velocity (turbulence, two phase flow, etc.) and derive related parameters, such as shear stress, flow angle, Reynolds number, velocity probability, turbulence spectra, viscosity,[29] etc. It is not at all surprising that many dozens of applications have been found in science and industry, and only a selection can be mentioned here. Many applications are of course mentioned in the books and conference proceedings concerning laser anemometry.[2-4,7,18,30-35,50]

Long range anemometry (50 m or more) has been performed using c.w. CO_2 lasers which emit eye safe 10·6 μm wavelength, infrared radiation. The monostatic heterodyne Doppler configuration described earlier has been used to measure crosswind and vortices, wakes and turbulence strength and duration behind aircraft of all sizes landing at airports; thus, safety during landing may be increased. Similar anemometers have been fitted into aircraft to determine true airspeed[36] and the possibility exists of simultaneous early detection of clear air turbulence, a major hazard to all aircraft in flight.[37] An entirely different application is the monitoring of effluent from industrial smoke stacks.[38]

Medium range anemometry (greater than say 1 m distance) usually employs visible lasers. Because of the range and collection solid angle of receiver optics, signals tend to be too low, particularly in backscattering, for processors other than the photon correlators to be used. Such situations are found when perhaps investigating super- and hypersonic flow inside large wind tunnels. Abbiss and his colleagues at Royal Aircraft Establishment Farnborough have pioneered the use of photon correlation anemometry under such conditions, obtaining impressive results (see ref. 7).

Similar pioneering work has been performed by Smart and his colleagues at Rolls Royce, Derby in studies on the RB211, Olympus 593 (Concorde) and other jet aeroengines.[39] Of great topical interest here is fuel conservation, thrust increase and noise reduction. It is of interest to note that all the world's major aerospace companies, Rolls Royce, Boeing, Pratt and Whitney, Grumman, NASA, etc., possess and use photon correlation anemometers for this type of work. Other similar investigations include studies of rocket exhausts[40] and helicopter rotor blade design.

Other medium range uses of anemometry include the study of cement powder manufacture in very hot 40 ft diameter drums, improvements in mixing and combustion efficiency of industrial gas burners (where pioneering work has been done by Birch, Brown and colleagues of the British Gas Corporation[41]), study of coolant flow inside nuclear reactors and spectacular measurements made from inside the bowels of a full-scale ship looking up to 7 m into the water flow around the screw.[18]

Work of equal importance to the above but perhaps less exotic is performed at short ranges, say 1 m or less. Petrol and diesel internal combustion engines,[42] turbines and turbo-compressors[31,49] have all been the subject of intensive investigation into their internal flow and combustion properties. An improvement in efficiency of just a few percent can reap great rewards for the manufacturers and considerably help fuel conservation.

Investigations of small bodies and models is also being actively pursued: the investigation of rockets, wing and aerofoil design effects on drag reduction, flow around ships' hulls and cavitation behind propeller blades,[7] and flows in canals, estuaries, rivers and around tall buildings in high winds for example.

Quality and quantity control in industrial processes is of great importance. Laser anemometry has been successfully used to measure accurately and control length[43] and flow rate of opaque liquids and solids, such as molten or sheet steel and plastic fibres.[34] In performing measurements on solid surfaces signal intensities may be far too high for detectors, and yet Doppler signal quality may be very poor due to having many scattering particles in the probe volume simultaneously. Furthermore, significant errors may be caused by using receiving optics with a lens to collect the scattered light! Such errors are only easily observed by observing a correlation function of the signal. The cause of the errors is due to curved wavefronts at the probe volume and the regime of incoherent detection inherent in using a wide aperture lens to collect light. The details are explained elsewhere,[44] but briefly the remedy is to employ the simplest receiver possible, using no lens but only a pinhole whose dimensions are defined by eqn (4). The arrangement is known as coherent detection and requires virtually no alignment, except to point the receiver approximately in the direction of the probe volume.

Further short range applications are quite diverse in nature, from studies of the mixing of liquid and gas fuels during combustion, to flow velocity in ocean currents,[45] from electric arcs and plasmas to investigations of fluid flow in pipes and ducts and around objects generally, and from the pneumatic transport of solids to measurement of the initial velocities of projectiles.

All of the applications mentioned so far have concerned motions from a few centimetres per second to many hundreds of metres per second. However, the laser anemometer is equally capable of measuring quite microscopic motions, for example airflow in model lungs,[46] study of the motion of highly viscous oils and greases,[29] vibration,[47] nerve and muscle movement and, surprisingly, plant growth which may be a few microns per minute. One particularly appealing application has been the measurement of blood flow in the retinal vessels of the human eye using a few tens of microwatts of laser light, yet achieving accurate velocities in 100 ms.[48] It is hoped that the technique will replace the lengthy and less accurate methods presently used involving dye and radioactive tracer injections.

Decreasing velocity leads eventually to an overlap with another laser technique, photon correlation spectroscopy (of say Brownian motion), which is fully dealt with elsewhere.[7] In the overlap area lie two notable applications, the study of electrophoretic transport[7] and the investigation of sperm motility to increase fertility.

It can be seen that laser anemometry has been applied successfully to many problems in engineering and science and there is no reason to doubt that its fields of application will continue to increase, particularly with the advent of fibre optic anemometry and its potential advantages.[50]

REFERENCES

1. Abbiss, J. B., Chubb, T. W. and Pike, E. R. *Opt. Laser Technol.*, **6** (1974), 249–61.
2. Cummins, H. Z. and Pike, E. R. *Photon Correlation and Light Beating Spectroscopy*, 1974, Plenum Press, New York.
3. Photon correlation techniques in fluid mechanics. *Physica Scripta*, **19**(3) (1979).
4. Durst, F., Melling, A. and Whitelaw, J. H. *Principles and Practise of Laser-Doppler Anemometry*, 1976, Academic Press, London.
5. Foord, R. *et al. J. Phys. D.*, **7** (1974), L36–L39.
6. Abbiss, J. B. and Mayo, W. T. *Applied Optics*, **20** (1981), 588–90.
7. Cummins, H. Z. and Pike, E. R. *Photon Correlation Spectroscopy and Velocimetry*, 1977, Plenum Press, New York.
8. Sonnenschein, C. M. and Horrigan, F. A. *Appl. Optics*, **10** (1971), 1600–4.
9. O'Shaughnessy, J. and Pomeroy, W. R. M. *Opt. Quant. Electr.*, **10** (1978), 270–2.

10. McLaughlin, D. K. and Tiederman, W. G. *Phys. Fluids*, **16** (1973), 2082.
11. Steenstrup, F. V. *Disa Inf.*, **18** (1975), 21–5.
12. Buchabe, P., George, W. K. and Lumley, J. L. *Ann. Rev. Fluid Mech.*, **11** (1979), 443–503.
13. Jackson, D. A. and Eggins, P. L., *AGARD-CP-193*, 1976.
14. Deighton, M. O. and Sayle, E. A. *Disa Inf.*, **12** (1971), 5–10.
15. Alldritt, M., Jones, R., Oliver, C. J. and Vaughan, J. M. *J. Phys. E.*, **11** (1978), 116–19.
16. Birch, A. D. and Dodson, M. G. *Optica Acta*, **27**(1) (1980).
17. Brown, R. G. W. and Gill, M. E. *Proc. ECOSA '82*, SPIE Vol. 369, paper 15, 1983.
18. Schultz Du Bois, E. O. (Ed.). *Photon Correlation Techniques in Fluid Mechanics*, 1983, Springer-Verlag, Berlin.
19. Brown, R. G. W. *et al. Physica Scripta*, **19** (1979), 365–8.
20. Tanner, L. T. and Thompson, D. H. *Proc. Symp. on Instrumentation and Data Processing for Industrial Aerodynamics*, 1968, National Physical Laboratory, London.
21. Schodl, R. *Thesis DLR-FB 77-6*, DFVLR Köln, 1977.
22. Schodl, R. *Trans. ASME, J. Fluid Eng.*, **102** (1980), 412.
23. Ross, M. M. *Optica Acta*, **27**(4) (1980), 511–28.
24. Brown, R. G. W. and Pike, E. R. *Opt. Laser Technol.*, **10** (1978), 317–19.
25. Mayo, W. T. *Optica Acta*, **27**(1) (1980), 53–66.
26. Greated, C. In: *The Engineering Uses of Coherent Optics*, ed. E. R. Robertson, 1976, Cambridge University Press, Cambridge.
27. Pusey, P. N. *J. Phys. D.*, **9** (1976), 1399–409.
28. Glass, M. and Kennedy, I. M. *Combust. and Flame*, **29** (1977), 333–5.
29. Jackson, D. A. and Bedborough, D. S. *J. Phys. D.*, **11** (1978), L135–L137.
30. *Proc. 2nd Int. Workshop on Laser Velocimetry*, 1974, Purdue University.
31. Laser optical measurement methods for aero engine research and development. *AGARD-LS-90*, 1977.
32. Applications of non-intrusive instrumentation in fluid flow research. *AGARD-CP-193*, 1976.
33. Durst, F. and Zarè, M. Bibliography of laser Doppler anemometry literature. *Disa Inf.*, 1974.
34. The accuracy of flow measurement by laser Doppler methods. *Proc. LDA Symposium*, 1975, Copenhagen.
35. *Proc. 4th Int. Conf. on Photon Correlation Techniques in Fluid Mechanics*, 1980, Stanford University.
36. Munoz, R. M., Mocker, H. W. and Koehler, L. *Appl. Optics*, **13** (1974), 2890–8.
37. Foord, R., Jones, R., Pomeroy, W. R. M., Vaughan, J. M. and Willetts, D. V. *Proc. SPIE*, 1979.
38. Brown, A., Thomas, E. I., Foord, R. and Vaughan, J. M. *J. Phys. D.*, **11** (1978), 137–45.
39. Smart, A. E. and Moore, C. J. *AIAA Journal*, **14** (1976), 363–70.
40. Kugler, H. P. *AGARD-LS-90*, 1976; *Physica Scripta*, **19**(3) (1979).
41. Birch, A. D., Brown, D. R., Dodson, M. G. and Thomas, J. R. *J. Phys. D.*, **8** (1975), L167–L170.

42. Cole, J. B. and Swords, M. D. *Applied Optics*, **18**(10) (1980), 1539–45.
43. Botcherby, S. C. L. and Bartley-Denniss, G. A. *Proc. Inst. Mech. Engrs*, **183** (1968–69), Pt 3D, 25–8.
44. Pike, E. R. In: *The Engineering Uses of Coherent Optics*, ed. E. R. Robertson, 1976, Cambridge University Press, Cambridge.
45. Stachnik, W. J. and Mayo, W. T. *Proc. Oceans '77*, 1977, 18A/1-5.
46. Greated, C., Durrani, T. S. and Ludlow, M. F. *Proc. Electro-Optics International*, 1974.
47. Simpson, D. G. and Lamb, D. G. S. *Nat. Eng. Lab. (UK) Report NEL-639*, 1977.
48. Hill, D. W., Young, S., Parker, P. and Pike, E. R. *IEEE J. Quant. Electr.*, **QE-13**(9) (1977).
49. Langdon, P. *High Speed Diesel Report*, **1**(5) (1982), 20–4.
50. *Proc. Int. Symp. on Applications of LDA to Fluid Mechanics*, 1982, Lisbon.

APPENDIX: LASER BEAM PROPERTIES

Kogelnik[1] has shown that a laser beam waist is only formed at the focal distance f, from a lens of focal length f, when a waist is present at the same focal distance behind the lens. However, if the waist is a large distance behind the lens, the waist formed by the lens may be extremely close to the focal plane of the lens.

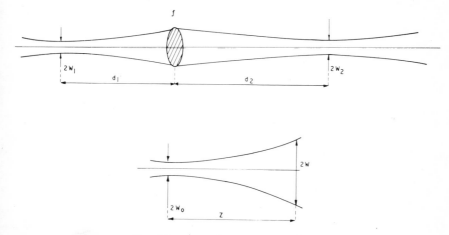

Fig. A1. Laser beam propagation.

Various equations are useful to calculate TEMoo laser beam parameters due to lenses. Referring to Fig. A1:

1. Position (d_2) of a beam waist after passage through a lens, given $d_1, f,$ W_1 and λ (W_1 = initial beam *radius*)

$$d_2 = \left\{ \frac{(d_1 - f)f^2}{(d_1 - f)^2 + \left[\frac{\pi(W_1)^2}{\lambda}\right]^2} \right\} + f$$

2. Waist *radius* (W_2) after passage through a lens, given W_1, d_1, f and λ

$$W_2 = \frac{W_1 f}{\left\{ (d_1 - f)^2 + \left(\frac{\pi(W_1)^2}{\lambda}\right)^2 \right\}^{1/2}}$$

3. Beam *radius* having travelled a distance Z from a waist of radius W_0

$$W = W_0 \left\{ 1 + \left(\frac{\lambda Z}{\pi W_0^2}\right)^2 \right\}^{1/2}$$

Note also that a laser beam should not be passed through an aperture of less than 1·7–2 times its e^{-2} diameter, or else it is significantly modified by diffraction effects.[2]

References

1. Kogelnik, H. *Bell Systems Tech. J.*, **44** (1965), 455–94.
2. Dickson, L. D. *Appl. Optics*, **9** (1970), 1854–61.

APPENDIX: BASIC PRINCIPLES OF OPTICS

A. R. LUXMOORE

Department of Civil Engineering, University College, Swansea, UK

1. INTRODUCTION

The purpose of the Appendix is to introduce readers having little familiarity with optics to some of the basic concepts and devices referred to in the book. It does not purport to be a rigorous introduction to optical theory, neither is it comprehensive, as there are many excellent textbooks in this area already. In addition to the basic concepts, it also contains a brief description of conventional optical instruments, i.e. image forming systems, that are well-established in engineering measurement.

The study of classical optics is conveniently divided into geometrical and physical optics. The former is restricted to the use of the laws of reflection and refraction in explaining the behaviour of optical components, and cannot explain interference and diffraction effects, which are covered by the latter. Geometrical optics, in the form of ray tracing, is used widely in the design of lenses and optical instruments.

Modern physics recognises the discrete nature of light in the form of photons, and quantum optics is important in explaining the interaction of light with matter, e.g. the photoelectric behaviour of materials. This has lead to a duality in explaining the physical behaviour of light, i.e. the particle nature and the wave nature. For our purpose, light will be considered as a wave phenomenon only, as this provides the simplest explanation of most physical optics phenomena.

267

2. THE NATURE AND PROPAGATION OF LIGHT

The wave nature of light was clearly demonstrated by Thomas Young at the end of the eighteenth century, but it was Maxwell, in 1873, who provided the unifying mathematical theory for the emanation of electromagnetic waves from an oscillating dipole. This theory identified light as part of the electromagnetic spectrum, and brought together a number of empirical laws governing the behaviour of the spectrum. The theory was embodied in four differential equations, governing simultaneous propagation of electric and magnetic radiation. In one dimension the solution of these equations can be represented as a travelling transverse wave, i.e. of the form

$$y = A_0 \sin\left(2\pi ft - \frac{2\pi x}{\lambda} + \Phi\right)$$

where f is the frequency, x the distance, λ the wavelength, Φ the phase and A_0 the amplitude (the Annex shows how a travelling wave may be represented by complex algebra). For a given point in space, we can simplify this disturbance to

$$y = A_0 \sin\left(2\pi ft + \Phi_1\right) \tag{1}$$

Alternatively, we can fix the instant of time, and so obtain the distribution of the disturbance in space

$$y = A_0 \sin\left(\Phi_2 - \frac{2\pi}{\lambda} x\right) \tag{2}$$

As the period for one cycle, $\tau = 1/f$, and the length of one cycle is the wavelength λ, then the velocity, c (= distance/time) = λf. The value of c for light in air was well established in Maxwell's day, and one convincing aspect of his theory was the independent prediction of c from the values of electric permittivity (dielectric constant) and magnetic permeability of free space.

Rigorous solution of the Maxwell equations show that the electric and magnetic vibrating vectors are mutually perpendicular and in phase. For the visible region of the electromagnetic spectrum, the magnetic vector can be ignored in many instances, as the interaction of light with a transparent medium is controlled by the electric vector and the dielectric constant of the medium. For example, the refraction of light at a boundary can be deduced from Maxwell's equations by applying the conditions that (a) the magnetic field crosses the boundary without change, and (b) the tangential component of electric force and normal component of electric displacement are continuous at the boundary (assuming that only the dielectric constant changes significantly at the boundary, which is the usual case for transparent materials). These conditions produce the familiar relation

$$\mu \sin i = \mu' \sin i' \tag{3}$$

where μ and μ' are the refractive indices of the two media, and i and i' are the angles of incidence and refraction, i.e. the direction of light to the boundary normal. Similarly the law of reflection can be obtained by putting $\mu = -\mu'$, giving

$$i = -i' \tag{4}$$

(The electric vector undergoes a phase change of $180°$ when it is reflected— this is of little consequence in geometrical optics, but very important in physical optics.)

When $\mu > \mu'$ in the equation, i.e. the light travels from an optically dense medium to a lighter medium, it is possible to find a value of i ($= i_c$) for which $\sin i'$ is 1, and $i' = 90°$. For $i > i_c$ (the critical angle) total internal reflection occurs, obeying eqn (4). For a glass/air boundary, this value is typically $\simeq 45°$. A large number of unsilvered prisms are available which make use of this property, being arranged so that i is about $45°$ or more at the reflecting surfaces.

The refractive index of a transparent material is given by

$$\mu = \frac{1}{\sqrt{\varepsilon}} = \frac{c}{V} \tag{5}$$

where ε is the dielectric constant of the material and V is the velocity of light in the material (c refers to the velocity of light *in vacuo*). When light passes through a boundary dividing materials of different dielectric constants, the frequency of vibration must remain constant, and so the change in velocity causes a change in the wavelength of the light wave, as well as refraction. This change in wavelength plays an important part in many optical phenomena.

The value of μ usually changes slightly with wavelength for most materials, and this explains why prisms split white light into the component wavelengths. This effect is known as dispersion.

3. LIGHT RAYS, WAVEFRONTS AND OPTICAL PATHS

A point source of light (e.g. a single oscillating dipole) in an isotropic medium will emit spherical waves and the electric vector of these waves at any point can be represented by

$$y = \frac{A \sin}{r} \left(2\pi f t - \frac{2\pi r}{\lambda} \right) \tag{6}$$

where r is the distance from the source to the point of interest. For a particular value of r, the electric vector at any point in that spherical surface has the same value, varying harmonically between $\pm A/r$. The *intensity* of light, I, is the rate of flow of energy across a unit area perpendicular to the direction of propagation of the waves, i.e. along a radius, and since the energy of a vibration is proportional to the square of the amplitude:

$$I \propto \left(\frac{A}{r}\right)^2 \propto \frac{1}{r^2} \qquad (7)$$

This is the well known inverse square law, which governs calculations of illumination levels for all sources which are small relative to the illumination range.

The loci of points having the same value of the electric vector is known as a *wavefront* and in this example it is spherical. The normals to a wavefront, representing the direction of energy propagation, are the light *rays* and the concepts of wavefronts and rays are widely used in geometric optics.

Spherical wavefronts are very common in optics and the effect of lenses is to change the curvature of a wavefront (Fig. 1(a)). Another common wavefront is the plane wave (Fig. 1(b)) usually known as either collimated or parallel light, and produced by the action of a lens on a spherical wave.

In optics, we are particularly interested in the number of wavelengths that a light ray takes to go from one point to another. If this path crosses several materials with different refractive indices, the velocity, and hence wavelength, will change. However the velocity V in materials is given by

$$\mu V = c = \lambda f \qquad (8)$$

and

$$V = \lambda_V f$$

where λ_V = wavelength corresponding to velocity V. Therefore, $\mu \lambda_V = \lambda$, the wavelength in free space. The number of wavelengths in a particular material is given by x/λ_V where x is the geometrical distance traversed by the light in that material. Hence the total number of wavelengths is given by

$$\sum \left(\frac{x}{\lambda_V}\right) = \frac{\sum \mu x}{\lambda} \qquad (9)$$

and the product μx is known as the optical path length, a measure of the number of wavelengths traversed along a ray regardless of the actual wavelength in the material. The wavefront should always be computed in terms of the optical path length.

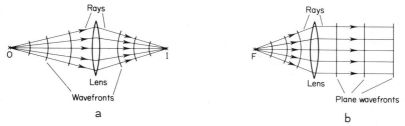

Fig. 1. Wavefronts with associated light rays. (a) Imaging of spherical wavefronts; (b) collimated light (plane wavefronts).

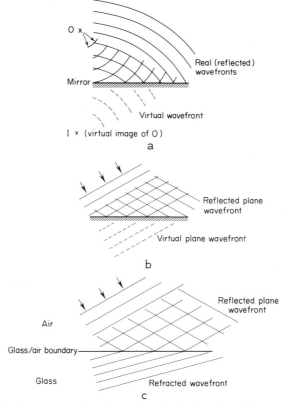

Fig. 2. Reflection and refraction of spherical and plane wavefronts. (a) Reflections of spherical wavefront; (b) reflection of plane wavefront; (c) combined refraction and reflection at a plane boundary.

We can now consider reflection and refraction in terms of wavefronts and rays. Figure 2(a) shows the reflection of a spherical wavefront from a flat mirror surface, and Fig. 2(b) the reflection of a plane wavefront. Figure 2(c) illustrates the combined reflection and refraction of a plane wavefront at the boundary of two transparent materials. The ratio of reflected intensity to refracted intensity is usually small for this situation and it can be predicted using Maxwell's equations.

A useful concept for explaining the geometrical propagation of light is Huyghen's principle. This supposes that every point on a wavefront behaves as a point source for further propagation of the wave known as secondary wavelets (Fig. 3(a)). The subsequent wavefront is then found as a combination of all the secondary wavelets. This principle accounts for the spread of light through a small aperture (which is why a single ray of light can never be isolated by a series of pinholes) (Fig. 3(b)). For large beams

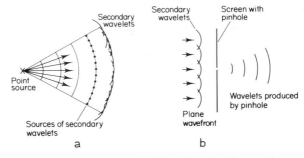

Fig. 3.   Light propagation by Huyghen's principle of secondary wavelets. (a) Spherical wave propagation; (b) plane wave propagation.

with plane wavefronts, the body of the beam will propagate in a rectilinear manner, but the light will always spread at the edge of the beam's aperture. This effect is known as diffraction, and will be discussed more fully in a subsequent section.

4.   POLARISATION

The oscillations of an atom or molecule which produces light can be arranged so that all the atoms or molecules vibrate either in the same plane, or in random directions. The former is achieved in a laser, which produces light waves of remarkable coherence and monochromaticity, as well as vibrating in a single plane, known as the plane of polarisation. All other light sources produce light waves vibrating in planes randomly aligned to the direction of propagation, known as unpolarised light. Unpolarised light can be polarised by passing through a polarising device, the most common nowadays being commercially available Polaroid sheet.

Unpolarised light can also be polarised by reflection. The boundary between two transparent media will produce both reflection and refraction (Fig. 2(c)), and Maxwell's equations predict the ratio of light transmitted to light reflected. Defining the coefficient of reflection, r, as the ratio of *amplitude* reflected to incident amplitude, the coefficient has two components: $r_\perp$ when the electric vector is perpendicular to the plane of incidence and $r_\parallel$ when it is parallel. These components can be expressed in terms of the angles of incidence and refraction i and i', as

$$r_\perp = -\frac{\sin(i - i')}{\sin(i + i')}$$

and

$$r_\parallel = \frac{\tan(i - i')}{\tan(i + i')} \tag{10}$$

When $(i + i') = \pi/2$, then $r_\parallel = 0$. Combining this condition with Snell's law (eqn (3)) gives

$$\frac{\mu'}{\mu} = \frac{\sin i}{\sin i'} = \frac{\sin i}{\sin(\pi/2 - i)} = \tan i \tag{11}$$

For a glass/air boundary this angle is $57°$, the well known Brewster angle for polarisation by reflection.

Polarisation by reflection at the boundary of a transparent medium, such as water, is a common occurrence in nature, so that a great deal of natural light is partially polarised (i.e. amplitudes of vibration in all directions at some point are not equal, but significantly reduced in some planes). This is the reason why polarising sunglasses significantly reduce the glare of sunshine from reflection by wet roads.

When light enters a transparent solid, its propagation is effected by the arrangement of the atoms and molecules. Solids with a random arrangement of atoms, such as glasses, are optically (as well as mechanically and electrically) isotropic, and the velocity of propagation is the same in all directions. Hence the refractive index is the same in all directions in these materials (this also follows from the fact that the materials' dielectric properties are isotropic). In liquids the constant but random movement of atoms also ensures that they are optically isotropic (glasses are, in effect, supercooled liquids) and gases behave similarly, although their refractive indices are almost unity, due to the wide separation of their atoms.

Any preferred orientation of the atoms in a transparent solid will produce optical anisotropy, and the velocity of light will vary in different directions. This is the case for many transparent crystalline substances (most solids have an ordered atomic arrangement, random arrangements are the exception) and the study of optical anisotropy can be used to elucidate the atomic structure of unknown crystals. In its most general form, this anisotropy can be represented by three orthogonal principal refractive indices, aligned with respect to the crystal axes, and the variation of refractive index with respect to these principal directions forms an ellipsoid figure, with principal indices μ_1, μ_2 and μ_3. This is the situation for a biaxial crystal, but some conditions of atomic symmetry give $\mu_2 = \mu_3$ and these are known as uniaxial crystals. Cubic crystals have $\mu_1 = \mu_2 = \mu_3$, and hence these crystals are optically isotropic, although not isotropic in the atomic sense.

Optical anisotropy not only affects the velocity of light, but it also produces double refraction, or birefringence. When light enters perpendicularly a slice of anisotropic crystal, the light ray is refracted into two components, and if the incident light is plane polarised, the two refracted components are also polarised in mutually orthogonal directions, aligned parallel to the direction of the principal indices of that slice. For a uniaxial crystal, one component will behave as if it were in an isotropic material, called the *ordinary* ray, whereas the other component does not, hence the *extraordinary* ray (the two components have different velocities). The refraction of the two components differs (Fig. 4) and hence two rays leave the crystal slice where only one entered. A common demonstration of this effect is the use of a thick slab of calcite, placed over a single dot on a piece of paper. An observer will see two dots through the calcite, and if the calcite is rotated in its own plane, one dot will remain stationary (the ordinary ray) and the other will rotate (the extraordinary ray).

Optically isotropic solids such as glass and plastics can be made slightly anisotropic or birefringent by the application of stress, which produces an orientating effect on the electric fields within the solids. This causes anisotropy of the dielectric constant and the refractive index. This stress birefringence is small compared to that produced by a natural crystal, but it can be demonstrated by arranging interference of the ordinary and extraordinary rays after suitable optical processing. The principal

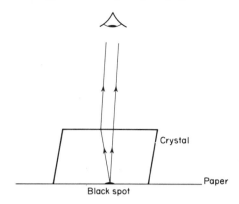

Fig. 4. Double refraction by calcite crystal.

refractive indices are aligned parallel to the principal stress axis, and the appliation of this phenomenon is known as photoelasticity, which is described in Chapter 2. Because of the smallness of the effect the difference in refraction is negligible and photoelasticity depends on the different velocities of the two polarised components to provide information about the stress components.

5. ABSORPTION

Solids and liquids will also absorb light, this absorption being very small in transparent materials. Absorption is usually expressed in terms of Lambert's law, where the intensity of light, I (which is proportional to the square of the amplitude), is reduced exponentially, i.e.

$$I = I_0 \exp(-kx) \tag{12}$$

where $x =$ distance traversed by light in medium and $k =$ absorption coefficient.

Some materials will absorb some wavelengths more strongly than others, hence the appearance of coloured objects in white light (sunlight). Also, excitation of atomic and molecular frequencies cause selective absorption, and the study of this phenomenon laid the experimental foundation for quantum physics. An important practical application of selective absorption occurs with polarised light in some crystals, known as 'dichroic' crystals. These absorb nearly all polarised light when aligned in one direction but if the plane of polarisation is rotated through 90°, a considerable proportion of light is transmitted. These dichroic crystals are the basis of Polaroid sheet, which contains ultramicroscopic dichroic crystals aligned in one direction (the alignment is achieved by stretching the sheets after incorporating the crystals).

In gases and liquids, light intensity can also be reduced by scattering, and this is discussed in the next section.

6. SCATTERING OF LIGHT

When a light beam encounters an obstacle in its path, there will be an interaction dependent on the size of the obstacle, or particle. For particles of molecular dimension, i.e. much smaller than the wavelength of light, there is a phenomenon equivalent to elastic collision, and some of the light intensity will be scattered in all directions. At this level, the scattered intensity is proportional to both the square of the volume of the particles, and also to $1/\lambda^4$ (Rayleigh scattering). Hence in white light, short wavelengths are scattered far more than the longer (red) wavelengths, and this is why the earth's atmosphere appears blue on a cloudless day.

As the particles become large compared with the wavelength, the scattering becomes very small and occurs only at the edge of the particle, effectively independent of wavelength. Dust particles in a room illuminated by sunlight will appear white whereas tobacco smoke (with molecular-sized particles) will appear blue. The tobacco smoke will truly scatter the light in all directions, albeit weakly, whilst the dust particles will reflect the light. Hence the term 'scattering' strictly applies to the effect of very small particles, but it is used generally to refer to the deviations of light by larger objects.

As an example, consider the mirror of Fig. 2. Its true surface will contain tiny asperities which will scatter a tiny amount of light in the true sense, but this is negligible compared with the specular reflection. If the surface is roughened on a scale that is large compared with the wavelength of light, there is no longer a specular reflection, and the light is scattered by the surface. In fact, this is really multiple random reflection and diffractions due to random orientations of the surface, but the beam of light is still effectively scattered by the rough surface. Hence it is important to distinguish between Rayleigh scattering and ordinary scattering.

7.   INTERFERENCE EFFECTS

If a beam of light is split into two equal components (a semi-reflecting mirror can do this) and then the two components recombine with a small phase difference between them, the light waves will add algebraically at every point of intervention to give a resultant wave. Hence if the two components, which are assumed to have equal amplitudes, are out of phase by 180° at any point, the waves will cancel at this point, yielding zero light intensity (Fig. 5(a)). If the components are in phase (or have differences corresponding to multiples of 360°) the waves will add to give twice their amplitude, and four times the intensity. Intermediate phase differences give intermediate intensities and this phenomenon is called interference. It is the basis of precision measurement using optical techniques, and enables displacement of fractions of wavelength (around 0·01 μm) to be measured.

The phenomenon of interference was used to establish the wave nature of

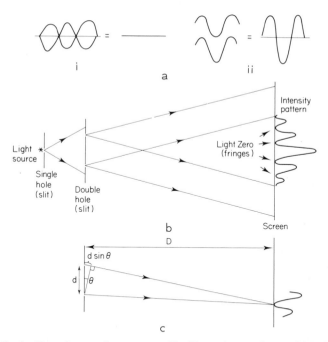

Fig. 5.   Optical interference demonstrated by Young's experiment. (a) Principle of (i) destructive and (ii) constructive optical interference; (b) Young's fringe experiment; (c) calculation of optical path difference for Young's experiment.

light in the 18th century by Young's experiment (Fig. 5(b)). Light spreading from a point source (which can be a small illuminated hole) is split into two components by two further small holes or slits placed close together. The light spreads from these small apertures and overlaps and, if a screen is placed in the region of overlap, bands of light and dark are observed in a direction perpendicular to the line joining the two holes. The dark bands correspond to points in the overlap where the two waves cancel, i.e. destructive interference, and vice versa from the light bands. The dark bands are called interference fringes, and the whole pattern is often referred to as an interference pattern or interferogram (the latter is used for a photograph of the pattern).

The phase differences causing interference in the Young's pattern arise from the differences in optical path length from any point on the screen to each hole. Considering two adjacent fringes, then the optical path difference must change by one wavelength. As the region of overlap is small, and fringes can only be observed when the holes are close together (distance d) and some distance ($= D$) from the screen (Fig. 5(c)) the optical path difference, OPD, for any point is approximately $a\theta$, but as $\theta \simeq x/D$, then

$$OPD = \frac{dx}{D}$$

For two adjacent fringes, x_1 and x_2 from the centre of the screen,

$$OPD = dx_1/D \qquad \text{and} \qquad OPD + \lambda = dx_2/D$$

Hence the fringe spacing $(x_2 - x_1)$ is given by $D\lambda/a$.

The optical arrangements for obtaining interference are many and various, and some of the arrangements useful for engineering measurements are described in the book. Nearly all of these methods use one light source, which is then split into two or more beams; apart from economy, there is a very good reason for this. Conventional light sources have a vast number of atoms giving out individual packets of light, which have a random phase relationship between each other. Any one of these packets can be split into components which can interfere with each other in a systematic manner, i.e. there is a simple geometrical relation between the crests and troughs at every point corresponding to a continuous single sinewave, but not with components from other packets (except in a random fashion). These millions of packets emitted will only interfere systematically by ensuring that their individual components combine, which means that any path length differences between components must not be longer than the individual packets. This is the concept of temporal coherence, and when the packets spread out, so that their components arise from different spatial volumes, as in Young's experiment, then there is also a need for spatial coherence.

The advent of the laser revolutionised interference experiments because light from each atom is emitted in phase and the laser cavity also ensures that light is emitted as a plane wavefront so that laser light is both temporally and spatially coherent. In fact one of the problems with laser illumination is preventing unwanted interference effects from stray reflections on the surfaces of optical and other components in the system. Interference effects can be obtained between two lasers, but this is not a practical tool and is not used in engineering measurements.

The splitting of a single beam into two components can be achieved in two ways: division of wavefront, illustrated by Young's experiment, requiring spatial coherence, a difficult proviso with conventional sources and division of amplitude, as might arise by splitting a beam with a semi-reflecting mirror so that half the amplitude is transmitted and the other half reflected.

An important proviso regarding interference is the need for the interfering components to be in the same plane of vibration. For incandescent light sources, the light is randomly polarised and there is no problem, but laser light is polarised (due to the Brewster effect at the windows enclosing the laser cavity) and subsequent reflections of the light can cause rotation of the polarisation plane and the possible alignment of two interfering beams with different polarisation planes. Interference will not occur in this situation, but can be obtained by passing the beams through a polarising medium, which will select components with the same polarisation plane.

8. DIFFRACTION OF LIGHT

8.1. Introduction

In Young's experiment, the light from the two holes were assumed to diverge sufficiently to overlap and hence produce interference. This phenomenon can be reproduced geometrically by using Huyghen's construction of secondary wavelets as it would not occur if the rectilinear propagation of light were strictly true. Hence at any opaque edge a small amount of light will spread around the edge and this is known as diffracted light.

The diffraction of light is a direct consequence of the wave nature of light, and is the limiting feature in any precise optical technique. The resolution of an accurately made lens is limited by the diffraction of light at its aperture. A laser cannot produce a perfectly collimated beam because of diffraction, and the beam will spread in diameter over long distances.

The diffraction of a spherical wave is referred to as Fresnel diffraction and that of plane wavefronts as Fraunhofer diffraction. The physical principles are the same, but Fraunhofer diffraction is much simpler mathematically, and most textbooks develop the theory of Fraunhofer diffraction first. This type of diffraction can be achieved in practice by having the illumination and observing points some distance from the diffracting object.

8.2. Fraunhofer Diffraction

The most fundamental diffracting object is a single slit. With a wide slit the light, in the form of a plane wavefront, passes through with a small spread at its edges (Fig. 6(a)). With a narrow slit, less than 0·1 mm wide, the spread of light becomes more significant, and the narrower the slit, the greater the spread (Fig. 6(b)). Although the mathematics is rather complicated, the amount of spread can be evaluated by calculating the positions of the first minima of the diffraction pattern (points A in Fig. 6(b)). This can be done very simply, using Fig. 6(c). Huyghens principle states that every point within the slit acts as a source of secondary wavelets, so that the light emitted in a given direction is obtained by summing the waves emitted by all the points at that angle, using the superposition principle. Alternatively, consider the light rays emitted from points A and B in Fig. 6(c), in the direction of the first minimum. Destructive interference occurs when their optical path difference is $\lambda/2$, but from the geometry of Fig. 6(c), this path difference is $(a \sin \theta)/2$. This path difference remains the same for all similar pairs of points, spaced apart by $a/2$, but moving down the slit, so that all points are paired. Hence the general condition for the position of the first minimum is

$$a \sin \theta = \lambda \tag{13}$$

Similar reasoning can be used for the other minima, giving, for order n,

$$a \sin \theta_n = (2n - 1)\lambda \tag{14}$$

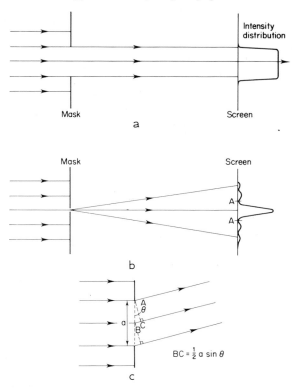

Fig. 6. Fraunhofer diffraction at single slit. (a) Transmission of light through large slit; (b) transmission of light through narrow slit; (c) optical path difference for single slit.

These equations show that as the slit width, a, becomes smaller, the angles for the minima become larger, so that narrower slits produce greater spread (the total light intensity, of course, is still reduced for narrower slits).

For the double slit, with a small spacing d between the two parallel slits, the interference between the two parallel slits, described previously, is superimposed on the basic pattern, as shown in Fig. 5(b). If a large number of equally spaced, parallel narrow slits is present in the mask, the combination of diffraction and interference produces separation of the diffracted light into distinct beams, or diffraction orders (Fig. 7), and the diffraction angle θ_n of the nth order is given by

$$d \sin \theta = n\lambda \tag{15}$$

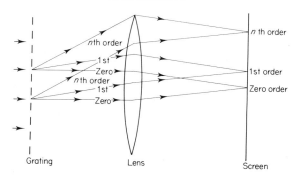

Fig. 7.   Fraunhofer diffraction by grating.

where d is the mean separation of the slits. The relative intensities of the
different orders depend on the mark/space ratio of the slits and bars
separating them, as well as on the nature of the diffraction grating. Slit and
bar gratings, more properly known as amplitude gratings, are only one
type, and similar gratings can be made using reflecting strips instead of slits,
or by introducing regular phase variations into the illuminating beam with
a corrugated transparent grating (Fig. 8) known appropriately as phase
gratings.

Diffraction gratings are used widely in astronomical and chemical
instruments for splitting optical signals into individual wavelengths (the
diffraction angle depends on the wavelength) and gratings with many
hundreds of lines per millimetre are used for this purpose. However,
diffraction affects gratings with as little as 10 lines mm^{-1} and in the
application of the moiré effect (Chapter 3), where gratings are superimposed, diffraction has to be considered to understand the resulting
signals. Below 10 lines mm^{-1} diffraction can be ignored, because the
individual orders are so close that they become indistinguishable.

The lens in Fig. 7 is used to focus the individual diffraction orders into
points of light in its back focal plane, so that they may be observed in a
convenient manner. Different orders can then be filtered out by placing an
opaque screen in this position and the desired orders transmitted through
suitable holes in the screen. This arrangement is known as a diffractometer
and is useful in some applications of the moiré effect.

Fig. 8.

Fig. 9. Airy disc (diffraction pattern from a pinhole).

Diffraction by a circular hole is qualitatively similar to that of a slit, the pattern being a series of concentric rings (Fig. 9) with the angular radius of the first minimum being given by

$$\sin \theta = 1 \cdot 22 \lambda / a \qquad (16)$$

where a is now the diameter of the pinhole. This pattern is known as the Airy disc and diffraction is only appreciable visually when a is very small, similar to the slit. On a macroscopic scale, however, diffraction by a large circular aperture places a diffraction limit on a lens. A lens viewing objects at a large distance will focus those objects in its back focal plane, as in Fig. 10. The smallest angular separation of two beams can only be resolved when the Airy discs formed by both beams are clearly separated, and Rayleigh suggested that this was just possible when the first minimum of one coincided with the maximum of the other (Fig. 10). Beams with angular separation less than this would not be resolved. As the angles are small, we can approximate $\sin \theta = \theta$ giving an angular resolving power of $1 \cdot 22 \lambda / a$. Hence lenses with large apertures, i.e. large a, have higher resolving powers,

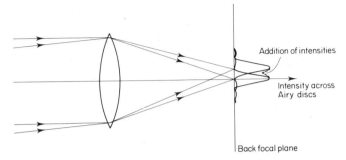

Fig. 10. Rayleigh diffraction limit of lens due to overlap of Airy discs.

and this is why it is so important for astronomical telescopes to have enormous apertures. Although diffraction limited lenses are readily available nowadays, resolution is often limited by poor measuring conditions and lens aberrations. These will be discussed later.

The Rayleigh condition can be used to deduce resolution limits for lenses that work with images outside their focal plane, e.g. magnifying lenses, yielding equations similar to $1.22\lambda/a$. Modern assessment of lenses uses the optical transfer function to determine resolution, as this takes into account the fact that resolution is not a simple cut-off, but contrast in microscopic images is progressively reduced as the detail gets finer until no contrast is available at the Rayleigh limit.

Diffraction is equally significant around small opaque objects, and this has provided a useful measuring tool for engineers. A photographic negative taken on a high resolution film with a large aperture camera lens (typically $f2$) of a scattering surface with microscopic scattering points, e.g. concrete, retroreflecting paint, etc., will contain very small images of the scattering points, and these images can be used to diffract a beam of light projected onto the negative over relatively large angular cones. If a double exposure is made on the same negative with the surface displaced by a small amount between the two exposures, the processed negative will diffract light from the two identical but displaced random patterns of microscopic scatterers, giving two overlapping cones which will interfere in a manner analogous to Young's experiment. These Young's fringes can be used to measure very small displacements of surface (or parts of surfaces) and the technique has been discussed in the book. The scattering effect can be produced by a grating stuck onto the surface (Chapter 3) or by illuminating any rough surface with laser light (Chapter 7).

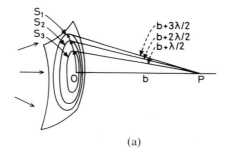

(a)

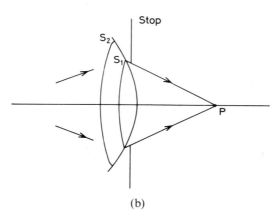

(b)

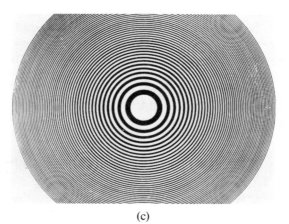

(c)

Fig. 11. Principle of Fresnel diffraction and Fresnel zone plate. (a) Spherical zones producing half wave differences at P. (b) Stopping out of higher order amplitudes. (c) Fresnel zone plate.

8.3. The Zone Plate

Although Fraunhofer diffraction can be used to explain most of the physical phenomena of diffraction, the action of a zone plate requires the use of Fresnel diffraction. A spherical wavefront (Fig. 11) can be divided into a series of half-wave zones, so that the secondary spherical wavelets originating in consecutive zones arrive at a point P, ahead of the wavefront with one-half wavelength differences. The distance from P to the border of each zone is one-half wavelength longer than the preceding zone. If a circular stop is placed between the wavefront and P, and the diameter of the stop allows only light from the first zone through (Fig. 11(b)), then the illumination at P will result from the amplitude of this zone. If the stop diameter is increased to allow light from the first two zones to pass, there is destructive interference and hence no illumination at point P. If the stop is increased incrementally to include more zones an alteration from dark to light illumination occurs at P.

A zone plate utilises the above construction to produce only constructive interference and hence bright illumination at P. Alternate zones are blocked out, starting with a transparent zone at the centre (Fig. 11(c)). As the zone plate is flat, the zones must allow for the curvature of the wavefront and this is done by making the radius of each progressively larger ring proportional to the square root of consecutive whole numbers. The zone plate will now act as a focusing device, i.e. a lens, and although not as efficient as a glass lens, it is very much easier to construct when devising single focusing devices with focal lengths of several metres (see Chapter 4). The focal length is adjusted by the absolute size of the central clear zone: the smaller the zone the shorter the focal length.

9.   BEHAVIOUR OF LENSES AND MIRRORS

9.1. Prisms

A glass prism (Fig. 12(a)) is used to deviate light by the principle of refraction, and this deviation increases as the angle, A, of the prism increases. Because shorter wavelengths (blue end of spectrum) are refracted more than the longer (red) wavelengths, increased deviation means increased dispersion of wavelengths, and prisms are also used in the same manner as gratings, to study the wavelengths in any light sample.

Prisms can also be used as reflectors, making use of total internal

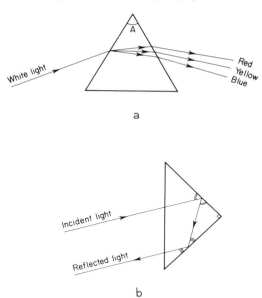

Fig. 12. Refraction and reflection by glass prisms. (a) Dispersion of white light into continuous spectrum of colours (only three shown for illustration); (b) right-angled prism used as retroreflector.

reflection (Fig. 12(b)) and it can be shown that light incident at any angle on the retroreflector of Fig. 12(b) will be returned in the same direction, the principle used in the 'cat's-eye' reflectors on major roads in the UK. Many complex prisms are manufactured nowadays for use in optical instruments, using both refraction and reflection to invert images, reduce the length of lightpaths, and achieve other desirable effects.

9.2. Simple Lenses

The refraction behaviour of a prism is the basic mechanism for the focusing action of a positive (convex) lens. The lens is effectively made up of a large number of elementary prisms with increasing deviation (Fig. 13) so that, as in Fig. 13(a), collimated light is brought to a focus (alternatively, a light source placed at this focal point will produce collimated light). The same lens will focus light from a point source, placed outside the focal distance, f, to an equivalent point on the other side of the lens (Fig. 13(b)), and the focal

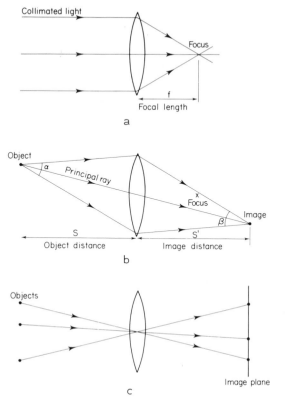

Fig. 13. Imaging by a thin convex lens. (a) Focusing of collimated light; (b) imaging of point source; (c) geometrical relation between object and image points.

points (or foci) can be used to establish this image point. Simple geometry shows that the relation between the object distance S and the image distance S' is given by

$$\frac{1}{S} + \frac{1}{S'} = \frac{1}{f} \tag{17}$$

The amount of light collected by a lens is measured by the solid angle α (Fig. 13(b)) which is a function of the lens aperture and distance S. For lenses viewing distant objects, the amount of light is measured by the solid angle β in the image plane (Fig. 13(b)) and as this is related to the lens aperture divided by the focal length, f, it is common practice to express the lens aperture as a ratio of the focal length. This is the basis of the 'f'-number used in cameras to control the lens aperture and hence the light gathering power of the lens. The ratio of f-numbers is chosen so as to increase the light gathering power by equal increments, but as this power is proportional to the area of the lens, and hence the square of the diameter, the standard f-numbers have the rather peculiar values 22, 16, 11, 8, 5·6, etc. The use of standard f-numbers simplifies the exposure requirements of different films, and film manufacturers usually quote exposures in terms of a given exposure time (usually 1/100 s) and the appropriate f-number.

The geometrical relation between object and image point is decided by the principal ray, which always passes through the centre of a simple lens (Fig. 13(c)) (images can be produced by allowing only the principal rays through a small hole, which is the principle of the camera obscura, but it produces a very weak image). The magnification, i.e. ratio of image to object size, is obviously equal to the ratio of image and object distance S'/S. Practical considerations limit the angular range over which a lens can be used, and modern camera lenses usually have this limit quoted.

Negative (concave) lenses will diverge a collimated beam (Fig. 14(a)) and cannot form a real image of a point (Fig. 14(b)). They are, however, used in conjunction with other lenses, e.g. the eye, and are said to produce a virtual image, indicated by the dashed lines in Fig. 14(b). Similarly, a convex lens will only produce a virtual image if the object point is between the lens and the focal point (Fig. 14(c)).

Simple lenses are manufactured with spherical surfaces and geometry shows that this will produce an adequate focusing effect if the 'f' number is fairly large, e.g. $f8$, corresponding to a small aperture. Apertures larger than this will not focus the outer rays to the same point as the inner, or paraxial rays, and the larger the aperture, the greater the error or aberration. In addition, for lenses used with polychromatic light, dispersion of the light will cause different colours to be focused at different points and so produce chromatic aberrations. This is also worse for large apertures where the deviation, and hence dispersion, is greatest.

The presence of aberrations reduces the resolving power of a lens, and for cheap lenses, aberrations are more serious than diffraction limits. Well corrected lenses, however, are often diffraction limited only. It is sometimes useful to consider the effect of geometric aberrations as perturbations of a wavefront (Fig. 15). The point source at a focus should produce ideally collimated light with a plane wavefront, but the wavefront will deviate from this ideal (Fig. 15(a)) especially away from the paraxial region. Similarly, an

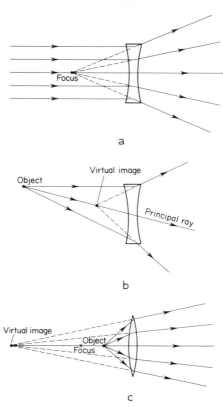

Fig. 14. Formation of virtual images by divergent rays. (a) Divergence of collimated light by negative lens; (b) virtual image formed by negative lens; (c) virtual image formed by positive lens.

incident spherical wave should be refracted into another spherical wave, but the aberration will also perturb this refracted wavefront (Fig. 15(b)).

9.3. Compound Lenses

The correction of aberrations is usually achieved by using several glass lenses (elements) combined together into one thick, or compound lens, enabling apertures of $f2$ to be achieved with very little residual geometric (Siedel) or chromatic aberrations. The latter are reduced by making use of glasses with different dispersive powers.

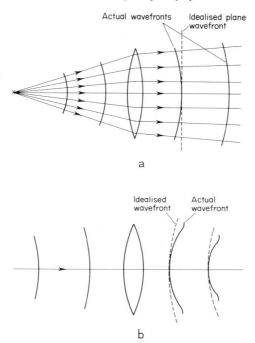

Actual wavefronts   Idealised plane wavefront

a

Idealised wavefront   Actual wavefront

b

Fig. 15.   Effect of lens geometric aberrations on wavefronts. (a) Distortion of plane wavefront; (b) distortion of spherical wavefronts.

Compound lenses behave in a similar manner to a simple lens, but the principal rays no longer pass through a single optical centre, but through two nodal points (Fig. 16) which are fixed points for that particular system. For some lenses, these points can exist outside the glass.

The aperture of a compound lens is not simply the size of the smallest element, and the position of aperture stops is very important as these limit the use of the lens to a given angle of view, as described previously. Compound lenses are designed for use at a given image, or field size, but for most lenses, reduction of the effective aperture will give reduction of the aberrations but with a consequent loss in light gathering power and resolution. Using compound lenses at distances and field sizes other than their design range will usually lead to poorer performance.

The simplest type of compound lens can be made by placing two simple lenses together to increase the power of the system. If f_1 and f_2 are the focal

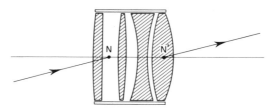

Fig. 16.   Compound lens with nodal positions.

lengths of the respective lenses, the focal length of the combination, f, is given by

$$\frac{1}{f} = \frac{1}{f_1} + \frac{1}{f_2}$$

(18)

Simple lenses (especially those used in ophthalmic work) usually have their dioptic power, D, quoted, where

$$D = \frac{1}{f} \qquad \text{with } f \text{ measured in metres}$$

(19)

Hence the combined power of two lenses (or more) in contact can be obtained by adding together their individual powers in dioptres.

9.4. Spherical Mirrors

The concave mirror in Fig. 17 has similar properties to a convex lens, as it can also focus an object point to an image. The focal length is half the radius of curvature, and although a mirror is not subject to chromatic aberrations (the laws of reflection are unaffected by wavelength), a mirror still suffers from geometric aberrations outside its paraxial region.

The best known use of concave mirrors is in astronomical telescopes, where very large diameter mirrors are used as the chief optical element. Smaller lens–mirror systems (catadioptic systems) have also been used to make compact, long focus telephoto lenses for cameras.

9.5. Compound Microscope

A simple magnifying glass can be held close to the eye to increase the power of the eye lens. Higher magnifications are achieved more conveniently by using two lenses (Fig. 18). The objective forms a real magnified image,

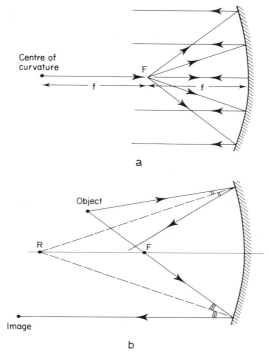

Centre of
curvature

F

f

f

a

Object

R

F

Image

b

Fig. 17. Uses of a concave mirror. (a) Collimated light; (b) imaging.

which is then magnified further by the eyepiece which forms an enlarged
virtual image viewable by the eye. Graticules and crosswires can be placed
in the image plane of the objective so that measurement can be made
without parallax (parallax occurs when the measuring scale is not
coincident with the object being measured). Magnifications up to $\times 1000$
are possible, but most engineering microscopes work in the $\times 10$– $\times 100$
range, with the objective being changed to give initial magnification of
$\times 2$– $\times 20$.

Engineering metrology often requires the accurate measurement of
profiles of fairly thick components, e.g. a screw thread. The use of a
conventional microscope can lead to serious errors (Fig. 19(a)) but a
toolmaker's (or profile) microscope avoids this problem by ensuring the
principal rays forming the image are parallel in the object space
(Fig. 19(b)). The effective aperture is in the back focal plane of the
microscope objective, and this is known as a telecentric stop. In practice,

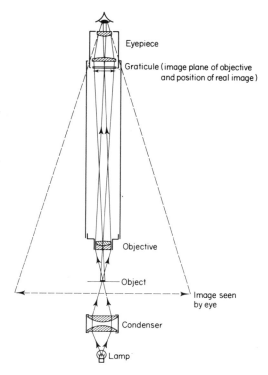

Fig. 18. Optical system of microscope.

collimated illumination is used, and the telecentric stop is not actually necessary. A similar principle applies to the use of profile projectors.

9.6. Telescopes

As a telescope is used to view very distant objects, they magnify the angular separation of objects (Fig. 20(a)). The objective lens forms a real image in its back focal plane, which is then magnified by the eyepiece forming a virtual image at infinity, increasing the angular magnification. Again, graticules and crosswires can be placed in the same plane as the real image for measurement purposes. This simple instrument produces an inverted image, and most terrestrial telescopes have additional lenses and/or prisms to produce a correctly aligned image. Focusing is obtained by moving the

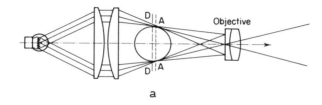

a

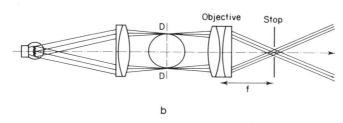

b

Fig. 19. Errors in profile measurement (a) using convergent illumination and (b) correction using collimated illumination. D–D, True diameter; A–A, apparent diameter.

eyepiece, and higher angular magnifications are obtained by using longer focus objectives.

Other telescope types are possible, the most common being the Galilean telescope (Fig. 20(b)). This produces a shorter instrument than the conventional astronomical telescope, its best known use being in opera glasses. The instrument is focused by moving the negative eyepiece lens. A combination of this principle with the astronomical telescope (Fig. 20(c)) produces the internal focusing telescope. Axial movement of the internal negative lens enables the real image to be focused on the graticule plane, where the eyepiece forms a virtual image at infinity. This provides a compact, robust instrument with no external moving parts to be damaged. It is widely used in theodolites and optical levels.

Telescopes are widely used for alignment, both in surveying and machine construction (see Chapter 4). The optical centres of the lenses define an optical axis and when focusing the telescope it is vital that there is no lateral displacement of the moving lens to alter this optical axis. A device which avoids this problem for measuring very small tilts is the autocollimator (Fig. 21). The semi-reflecting mirror allows the crosswires to be projected through the objective which is focused at infinity. A mirror is placed on the

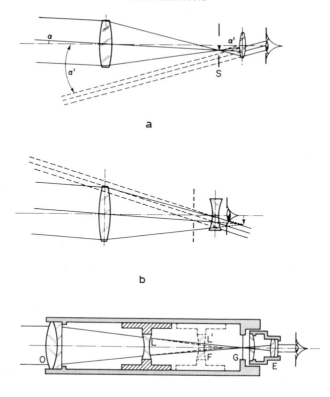

a

b

c

Fig. 20. (a) Astronomical telescope; (b) Galilean telescope; (c) internal focusing telescope. (Courtesy of Pitman & Sons, Ltd.)

engineering component and reflects the light back into the autocollimator, where an image of the crosswires can be seen by the eyepiece in the same plane as the actual crosswires. Measuring the difference in position of the image from the crosswires, by means of a graticule or micrometer, enables small tilts in the mirror to be determined.

10. THE DOPPLER EFFECT

When a high speed train gives a warning whistle passing through a station, people on the platform are aware of substantial lowering in pitch as the

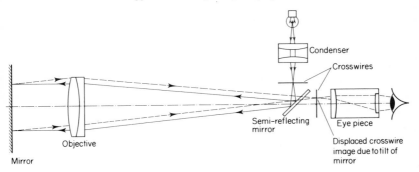

Fig. 21. Principle of autocollimator for measuring small tilts.

train passes. This is because the whistle is moving rapidly towards the people at first, thus compressing the sound waves and effectively increasing the frequency, and later it is moving away, thus stretching out the wave motion and lowering the frequency. This is the Doppler effect, and it is also present with light waves, except that the speed of motion must be very high in order to obtain any noticeable effect. The most famous application in optics is the red shift of galaxies moving away from our own galaxy at high speed. Less well known is the line broadening, i.e. small spread of frequencies, in a gas laser due to the motion of its atom. Application of the Doppler effect to anemometry is given in Chapter 8.

11. PROPERTIES OF THE GAS LASER

Reference has already been made to the laser, which is an acronym derived from Light Amplification by Stimulated Emission of Radiation. The device appears frequently in the book, and so a few facts about lasers will be summarised here.

The mechanism of stimulated emission cannot be explained by classical optics, but suffice to say that stimulated emission only occurs when a large number of excited atoms absorb and retransmit a given wavelength. This can be done by holding the atoms in a cavity which reflects the radiation many, many times, so that it stimulates further emission. The cavity can be a transparent solid with polished ends, or a tube containing a suitable gas mixture, such as the helium–neon laser. The latter is one of the most widely used lasers, and gas lasers have the general advantage that they emit continuously, i.e. continuous wave (c.w.) lasers, whereas solid lasers, such as synthetic ruby are pulsed. As gas lasers are the most widely applied in this book, further discussion is limited to this type.

The gas is held in a narrow bore tube with glass windows at either end, outside of which are two mirrors. In principle these could both be plane mirrors, but in practice alignment is much easier if one or both are slightly concave. The mirrors are multilayer with a reflectivity in the desired wavelength of 99 % hence allowing only 1 % of the beam intensity to be used. This is essential if stimulated emission is to be maintained and the selective nature of the mirrors suppresses other possible lasing wavelengths.

The reflectivity of ordinary glass windows at the end of the laser cavity is too high for a flat-ended laser to work, because the reflections prevent all the light from leaving the tube so that it can be reflected back *precisely* by the aligned mirrors. The laser uses the Brewster effect, discussed earlier, to overcome this problem. Light incident at the Brewster angle has a component of light with its polarisation plane parallel to the surface, passing through with no reflection, whereas the perpendicular component has a significant amount reflected. By arranging the laser windows to be set at the Brewster angle to the tube (approximately 57°) the parallel component passes through without loss, and it is the only component that lases, the other component dying away rapidly after a few passes. The polarisation plane inherent in a gas laser is thus related to the position of the tube window.

Stimulated emission is only maintained for the light passing up and down the tube, and any light emitted at a significant angle will die away. However, a wave travelling only slightly off-axis can zig-zag between the mirrors a sufficient number of times to lase, especially with curved mirrors, and a number of distinct transverse lasing modes are possible (called TEM, for 'transverse electromagnetic'). The different modes can be produced by adjusting the mirror design, but most lasers are designed to operate in TEMoo (by analogy with microwave usage) or *uniphase* mode, with the beam emerging from the laser as a single symmetric beam of light with maximum intensity at the centre, and falling away at the edges of the beam. Misalignment or dirt on the mirrors can produce other modes, which appear in cross-section as several beams, but for use in interference experiments and holography the laser must be adjusted for TEMoo. This is the only mode producing spatial coherence across the beam section.

The intensity distribution across the TEMoo mode is Gaussian, and is expressed by the following equation:

$$I(r) = I_0 \exp\left(-2r/w^2\right)$$

where I_0 is the intensity at the centre of the beam, r is the radial distance from the centre and w is the value of r for which the intensity has fallen to e^{-2} of its value at the centre, sometimes referred to as the e^{-2} points, and taken as the effective edge of the beam. Using the e^{-2} intensity points to define the beam, Fig. 22 shows how the beam spreads as it propagates in space. The minimum value of w ($= w_0$) is the position of a true plane wavefront, and in some lasers the tube is shortened by placing a plane mirror at this point. The plane wavefront can be reproduced outside the cavity by using a convex lens (Fig. 22) because the basic angular spread of the beam is still reproduced. This angular spread is very small, typically 0·5 milliradians, which is smaller than the spread due to diffraction of a plane wavefront of equivalent size.

The laser cavity is a resonator, with the light forming standing waves and the mirrors at nodal points. For a simple cavity with plane mirrors, stationary waves satisfy the relation

$$f = nc/2L$$

where L is the length of the cavity, and n is an integer. Different values of n produce different axial modes, and the difference in frequency, Δf, between two axial modes is given by

$$\Delta f = c/2L$$

In practice, the stimulated atoms will emit a small spread of wavelengths due to the movement and collisions between individual atoms, and hence it is possible for several different axial modes to be present in a laser output. This reduces the coherence length of the laser and should be limited as far as possible. Various methods are available for achieving this end.

In addition to multiple axial modes the laser cavity length will also vary due to temperature changes, vibration and similar effects. This causes an overall spread in the emitted wavelength for each axial mode, and for very

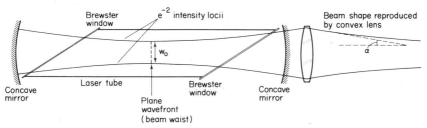

Fig. 22. Propagation of a c.w. laser beam (schematic). Beam divergence $= \alpha$.

precise applications these random changes should be corrected. This can be done by mounting one mirror on a piezoelectric crystal and using a servo-system to maintain an accurate wavelength. Such systems are available commercially.

Although the laser produces light with far greater temporal and spatial coherence than any other source, the spread in wavelength due to the reasons outlined previously can cause a sufficient loss of coherence to impair applications such as interference measurements and holography unless some simple precautions are taken. Different optical path lengths used to produce interference effects should be approximately the same length, or at least different by a distance equal to some multiple of the laser cavity. Working with distances that are multiples of the cavity length ensures the best coherent conditions, as all axial modes must have nodal points at the mirrors, and this correspondence is reproduced in space at distances equal to the laser cavity. Where this is not possible, stabilised lasers must be used.

BIBLIOGRAPHY

Beesley, M. J. *Lasers and their Applications*, 1971, Taylor & Francis, London.
Brown, E. B. *Modern Optics*, 1965, Reinhold Publishing Corp., New York.
Habell, K. J. and Cox, A. *Engineering Optics*, 1966, Pitman, London.
Hecht, E. *Theory and Problems of Optics*, Schaum's Outline Series, 1975, McGraw-Hill, New York.
Longhurst, R. S. *Geometrical and Physical Optics*, 1957, Longmans, London.
Open University. *The Wave Nature of Light*, Science Foundation Course Unit 28, 1971.

ANNEX: REPRESENTATION OF WAVE MOTION BY COMPLEX ALGEBRA

A complex number, z, has the form

$$z = x + iy$$

where $i = \sqrt{-1}$, and x and y are the *real* and *imaginary* components, respectively. As real and imaginary parts are quite separate, complex numbers can be represented on a Cartesian graph (the Argand diagram) (Fig. A1). From this diagram, we see that z can be rewritten as

$$z = A(\cos \theta + i \sin \theta)$$

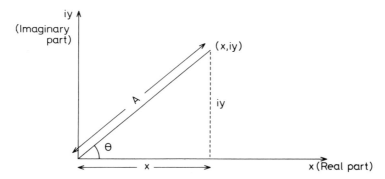

Fig. A1. Representation of complex number $z = x + iy$ by Argand diagram.

using $x = A \cos \theta$ and $y = A \sin \theta$. Euler's formula states that

$$\exp(i\theta) = \cos \theta + i \sin \theta$$

so that we can write

$$z = A \exp(i\theta)$$

where A is the magnitude and θ the phase of the complex number. The *complex conjugate* z^* is obtained by changing the sign of i wherever it appears in z, hence

$$zz^* = (A \exp(i\theta))(A \exp(-i\theta)) = A^2$$

$$A = \sqrt{zz^*}$$

Any cosine or sine can be represented by a complex exponential, by taking either the real or imaginary part to represent the appropriate component. The complex exponentials can be manipulated very easily in any calculation and then the original cosine or sine form obtained by taking the real or imaginary part of the answer. For wave motion,

$$\theta = 2\pi f t - \frac{2\pi x}{\lambda} + \Phi$$

and the sine wave becomes the imaginary part of

$$A \exp\{i[2\pi f t - (2\pi x/\lambda) + \Phi]\}.$$

In most calculations, it is only necessary to calculate the resulting amplitude, and this is obtained by using the complex conjugate as above.

This form of mathematical treatment is very useful in optical calculations, especially in connection with interference effects and holography (see Appendix in Chapter 6).

INDEX

303